OFFICE ERGONOMICS

Ease and Efficiency at Work

SECOND EDITION

OFFICE ERGONOMICS

Ease and Efficiency at Work

SECOND EDITION

Anne D. Kroemer
Karl H.E. Kroemer

CRC Press
Taylor & Francis Group
Boca Raton London New York

CRC Press is an imprint of the
Taylor & Francis Group, an **informa** business

First published in paperback 2024

First published 2017 by CRC Press
2385 NW Executive Center Drive, Suite 320, Boca Raton FL 33431

and by CRC Press
4 Park Square, Milton Park, Abingdon, Oxon, OX14 4RN

First issued in hardback 2019

CRC Press is an imprint of Taylor & Francis Group, LLC

© 2017, 2019, 2024 Taylor & Francis Group, LLC

Library of Congress Cataloging-in-Publication Data

Names: Kroemer, K. H. E., 1933- author. | Kroemer, Anne D., author.
Title: Office ergonomics : ease and efficiency at work / Anne D. Kroemer and Karl H.E. Kroemer.
Description: Second edition. | Boca Raton [Florida] : Taylor & Francis, CRC Press, 2016. | Earlier edition by Karl H.E. Kroemer and Anne D. Kroemer. | Includes bibliographical references and index.
Identifiers: LCCN 2016003235 | ISBN 9781498747943 (alk. paper)
Subjects: LCSH: Human engineering. | Work environment.
Classification: LCC TA166 .K78 2016 | DDC 620.8/2--dc23
LC record available at http://lccn.loc.gov/2016003235

ISBN: 978-1-4987-4794-3 (hbk)
ISBN: 978-1-03-292379-6 (pbk)
ISBN: 978-1-315-36860-3 (ebk)

DOI: 10.1201/9781315368603

Visit the Taylor & Francis Web site at
http://www.taylorandfrancis.com

and the CRC Press Web site at
http://www.crcpress.com

For Hiltrud; for Ellie

Contents

Preface

The reasonable man adapts himself to the world: the unreasonable one persists in trying to adapt the world to himself. Therefore all progress depends on the unreasonable man.

—**George Bernard Shaw,** *Man and Superman*

WORKING AT EASE AND EFFICIENTLY

We have written this book so that everyone can use it; anyone who uses existing or virtual keyboards, checks e-mails, makes calls, sends texts, sits, stands, writes, reads, tweets, posts, pins, snapchats, discusses, evaluates, prepares, presents, and makes decisions can find it useful. It is a practical book based on sound theory and research, augmented by our own experiences in offices. We based this book on the assumption that we all strive for some amount of productivity and success, and we would like to do so with minimal aches, pains, and discomfort. We would also like to have solid working relationships with colleagues and clients, suppliers and vendors, and superiors and subordinates. In the best of all worlds, we would also truly enjoy our jobs and happily lose ourselves in our work the way we lost ourselves in play as young children or lose ourselves in a riveting book or a scintillating movie as adults.

This book suggests how best to achieve this harmonious work scenario by optimizing the "fit" between the person—including his/her capabilities and limitations—the equipment used, and the person's environment. This, in a nutshell, is what ergonomics is all about. Accordingly, this book offers advice on how to set up the office, at home or at a company; how to select and arrange equipment and furniture; how to organize and pace work; how to understand and interact with colleagues and associates; and how to do all of this with appropriate ergonomic design to prevent repetitive strain injuries, musculoskeletal disorders, and the resulting long-term disabilities that can develop over time. In other words, we hope to show how to perform "at ease and efficiently"—the motto of ergonomics.

Given that it focuses on the "fit" between the person, equipment, and environment, ergonomics borrows from a number of disciplines, including industrial psychology, industrial design, industrial engineering, physiology, cognitive psychology, biomechanics, mechanical engineering, and anthropometry. You will notice this as we progress through the ensuing chapters.

DISCLAIMER: PEOPLE ARE DIFFERENT!

How boring if we were all the same! And yet that leads us to a small disclaimer. The recommendations in this book apply mostly to American, Down Under, and European users. For example, the chair measures suggested here fit their bodies, while other populations with fundamentally different body sizes would need other ranges in furniture dimensions and adjustments. Furthermore, working postures and habits can differ strongly among the various populations on earth. Preferences for colors, lighting, or temperature in the office vary from one continent and climate zone to the next. However, with modifications and adjustments as needed, these ergonomic principles apply everywhere.

HOW TO USE THIS BOOK

Please use this book in the manner that is best for you:

1. Read it straight through from beginning to end, as in a university course. Then consider whether you should change things in your office environment to be more comfortable.
2. Read a section of interest, think about applying that information, and proceed to the ergonomic recommendations that make use of that information.
3. Use the Index to look up a topic of concern to you, and proceed to the ergonomic recommendations that make use of that information.

DO LET US KNOW

If you find that this book is of help, or that we should include something else you feel is important, please reach out to Taylor & Francis Group at www.taylorandfrancis.com, or e-mail us directly at Anne.OfficeErgonomics@gmail.com.

We welcome your feedback!

Anne D. Kroemer Hoffman
Karl H.E. Kroemer

Authors

Anne D. Kroemer Hoffman received her MBA from the University of Virginia, Charlottesville, after which she moved into the world of advertising, holding progressively important roles in client service with world-class advertising agencies including Leo Burnett and J. Walter Thompson. She then left to pursue human rights interests by working with a large Chicago-based social service agency as director of marketing and development, where she developed and wrote all marketing brochures and literature, wrote press releases, authored and designed newsletters and annual reports, created and designed/wrote all promotional and volunteer recruitment materials, and handled all fundraising. While raising a young child, she worked in a community-based organization as director of outreach and marketing. Specific duties involved monthly e-newsletters, writing and updating an of-the-moment website, all marketing and public relations activities for designated areas, managing and marketing for two momentous annual events, and all social media, which she launched. She is currently employed as a freelance marketing and communications consultant in Chicago.

Please contact Anne Kroemer at anne.officeergonomics@gmail.com.

Karl H.E. Kroemer received his training as a mechanical engineer at Technical University Hannover, Germany: BS (Vor-Diplom), 1957; MS (Dipl.-Ing.), 1960; PhD (Dr.-Ing.), 1965. He worked as a research engineer at Max Planck Institute for Work Physiology in Germany for several years before relocating to the United States to take a position as research industrial engineer in the Human Engineering Division of the Aerospace Medical Research Laboratory at Wright-Patterson Air Force Base in Ohio. After accepting a job as director and professor in the Divisions of Ergonomics and Occupational Medicine at the Federal Institute of Occupational Safety and Accident Research, he returned to Dortmund, Germany, for several years. In his last intercontinental relocation, he took on the position of professor of ergonomics and industrial engineering at Wayne State University in Detroit, Michigan, for several years before settling in Virginia, where he became emeritus of industrial and systems engineering and director of the Industrial Ergonomics Laboratory at Virginia Tech in Blacksburg. Recently retired, he still writes and publishes and authors/coauthors more than 200 publications.

Please contact Dr. Kroemer at kroemer@vt.edu.

1 Office Ergonomics Defined

My office is at Yankee Stadium. Yes, dreams do come true.

—Derek Jeter, former MLB star for the New York Yankees

OVERVIEW

Office design consists of macro- and microergonomics; it encompasses size-able projects like the design of entire office buildings, medium projects like laying out individual workstations, and smaller-scale projects like devising and placing work station–related components—chairs, desks, computers, and even your computer mouse.

The person trained for this design work is an ergonomist, and it is his or her goal to arrange the office so that it

- Enhances people's well-being, both physically and cognitively
- Streamlines work
- Allows people to perform efficiently and safely

Over the past decade, office design has changed dramatically. The term "office" itself can now be as easily described as a traditional bricks-and-mortar building—with cubicles, office plants, an executive floor with top-to-bottom windows, and a 9 to 5 work environment—as it can be seen in a local coffee shop where an individual employee is telecommuting with the help of a laptop or a cell phone and/or having a tasty latte. We will address these developments in this chapter.

ERGONOMICS DEFINED

In the introduction to their 2001 book *Ergonomics: How to Design for Ease and Efficiency*, the Kroemers gave the following definition: "ergonomics is the application of scientific principles, methods, and data drawn from a variety of disciplines to the development of engineering systems in which people play a significant role. Among the basic disciplines are psychology, cognitive science, physiology, biomechanics, applied physical anthropometry, and industrial systems engineering." The components of ergonomics are pictured in Figure 1.1.

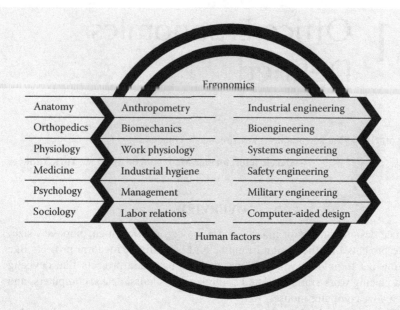

FIGURE 1.1 Components of ergonomics. (Adapted from Kroemer et al. 2001.)

Often called human factors engineering in the United States, ergonomics is neutral: it takes no sides—neither employers' nor employees'. It is not for or against progress; it is not a philosophy; it is not an opinion or a trend. Instead, it is a scientific discipline and technology, one that designs processes, products, and systems with the focus on the interaction between them and the people who use them. Briefly put, ergonomics designs machines, systems, and environments around the human, instead of making the human adapt himself or herself to the machine.

Distinctions are often made between "macroergonomics" and "microergonomics," although labeling human engineering tasks as one or the other can be a simple matter of judgment. Designing a public transportation system, for example, or a manufacturing plant, meaning a high and complex level of design, would fall under macroergonomics, as would sketching out the organizational setup of a company, with the concurrent technologies, employee interactions, and other socio-technological considerations. On the other hand, making a pencil sharpener or a toothbrush more user friendly and simple to use would fall under the definition of microergonomics. Of course, even designing a light pen or a computer mouse involves more than just considering hand size and mobility, and so to label human engineering tasks, micro- or macroergonomics might just be semantics.

Office ergonomics, with today's massive changes in how business is conducted, now moves way beyond the traditional office. Consider that some companies have abandoned bricks-and-mortar office buildings altogether; a company that

conducts all of its business online may eschew rent/lease, building maintenance, and other costs and have all employees telecommute. The benefits to a
telecommuter are clear—no lengthy and frustrating commutes, flexible work
schedules, a seriously relaxed dress code, a comfortable couch or lounge chair
and office desk. The ergonomic challenge, however, now becomes teaching the
remote worker how to apply ergonomics to his or her work station—whether it
be a home office, a coffee shop, or a poolside lounge chair—and reap the benefits of a safe, efficient, and pleasant work station.

Ergonomics applied to the whole organization directly or indirectly affects
every employee. Sure, there is the size and layout of our workspace to consider, but there is also climate control (heating and cooling), lighting and seeing, hearing and sound, design and comfort of workspace components like
chair and keyboard. In addition, organizational behavior—how people, individuals, and groups act in organizations—plays an important role in overall
office ergonomics; all of these topics are covered in further detail in Chapters 2
through 10.

Success of ergonomic effort is measured by improved productivity, efficiency, safety, and acceptance of the resultant system design and—last, but
truly not least—by improved quality of the human life. The latter, when you
consider it, is an enormous result indeed.

HUMANS RUN THE OFFICE

Unlike an automated factory, for example, the office is a work system that depends
entirely on humans: without them, no work is accomplished. Consequently, ergonomics focuses on the human as the most important component of the office and
adapts the office to the people involved. This human-centered design requires
knowledge of the characteristics of the people in the office—their physical dimensions, capabilities and limitations, and preferences.

Observing and studying office workers in their environment is not new; in the
early 1700s, Bernardino Ramazzini—often called the "father of occupational
medicine"—noted that workers who sat still, stooped, looking down at their work
often became round-shouldered and suffered from numbness in their legs, lameness, and sciatica. Ramazzini also took note of the debilitating writer's cramps from
which scribes often suffered in the office (Wright's 1993 translation of the 1713
book, pp. 180–185).

Ergonomics is not a "new" science, but rather one that relies on more than a
hundred years of physiological, psychological, and engineering observation and
research—even far longer if you consider Ramazzini's early efforts. Ramazzini
generalized that "all sedentary workers suffer from lumbago," and he advised that
people not sit still but instead move the body and "take physical exercise, at any
rate on holidays." He appears to have been prescient vis-à-vis the importance of
moving around, although he did not take it far enough; sitting for long periods

of time is associated with a number of risk factors. Higher cholesterol levels, larger waist lines, cardiovascular disease, and metabolic disease are chief among those myriad risk factors.

Side note: if you are interested in the debut and evolution of ergonomics as a science, we have put together an appendix in this book that will walk you through the colorful history of the science (Appendix A).

Today, musculoskeletal disorders in the hands and back, often together with eyestrain, are common complaints of people who operate computers in the office. With the amount of keyboarding and texting that goes on now, hand, wrist, and shoulder disorders are a fast-growing source of disability in the American workplace. This development is disappointing, even deplorable, because the extent of these disorders could be minimized by timely and proper application of ergonomic knowledge. Applying what we know today should help us avoid many of these problems in the future.

In modern times, many work-related musculoskeletal disorders are computer-related injuries: for example, the widely known disease "carpal tunnel syndrome" is often associated with keyboard use. However, while those who use electronics are subject to those injuries, even those who do not often suffer from work-related discomfort, pain, and disease may experience

- Lack of whole-body movement
- Poor or improper postures, especially when maintained for long periods, often caused by ill-fitting furniture including chairs
- Physical overexertion of hand, arm, and shoulder by repetitive work

Other problems are related to

- Inadequate lighting
- Excessive noise
- Stressful climate in the office both on physical and psychological aspects

We would like you to note here that a stressful office climate might not *seem* like a cause of physical disorders, but indeed it can be. Office conditions like the power structure within the whole company, working relationships between managers and employees as well as among employees, levels of work autonomy and responsibility, job security, and other organizational aspects can have overt negative effects on individuals (organizational behavior is covered in some detail in Chapter 2). Possible outcomes include lack of motivation, uncooperativeness, dissatisfaction with work, unhappiness in general, and even outright rage. The mind–body link means that these psychological effects aggravate physiological problems like the ones listed earlier. Physiological and psychological traits are often treated as separate academic topics, but in reality, they are inseparably intertwined and together determine a person's overall well-being.

Negative outcomes like those just mentioned are pressing issues. Work injuries, unhappiness, and discomfort are life-altering consequences that can profoundly affect an individual's overall well-being. Fortunately, they are avoidable, if we understand how to set up our "offices" (wherever those may be) to be comfortable and efficient.

The first step in setting up a work station is to understand that people do not come in one size, they do not have the same body proportions, and one (chair) size does not fit all. Moreover, we do not all have the same preferences and dislikes. We are unique, we behave differently, and not only do we *want* to be treated as individuals, we *need* to.

COMPUTERS CHANGED THE OFFICE IN THE 1980s

Computers are something we take for granted nowadays; it is almost impossible to visualize an office without one. But computer-free days did exist, and until the 1950s, repetitive work like mathematics calculations had to be done manually (or by abacus). When computers first made their debut, the earliest versions were developed and used as high-speed calculating machines. In fact, the first few decades of computers in business focused primarily on automating various industrial processes. It took another 10 years before the concept of using computers as tools to improve business productivity was realized.

In the 1980s, when personal computers began to proliferate in offices, analysts began to foresee the dramatic rise in productivity that technology might allow. But others took a grimmer view of the burgeoning use of PCs. Etienne Grandjean (1987), for one, a Swiss professor and researcher in ergonomics and work physiology, was apprehensive about the impact of the newly popular computer on the office worker. He was particularly concerned with the following:

- Potentially harmful radiation coming from the display, especially if the user were pregnant.
- Musculoskeletal overuse disorders due to repetitive activities, especially keyboarding. This had already occurred in the so-called repetitive stress injury (RSI) epidemic in Australia in the early 1980s.*
- Luminance of electronic displays in combination with lighting of the office.
- Layout of the computerized workplace for appropriate posture of the operator. Remember that at this point early computers had been simply plopped onto an existing desk or table originally designed for traditional office work.

While the fear of radiation damage subsequently diminished, Grandjean's other concerns were still very much in play in 1990 when Sauter and his coauthors addressed the issues of productivity and operator health in the computerized office. Office managers and designers had not foreseen that the computer would develop so

---

* The Australian RSI epidemic refers to an outbreak of work-related arm and hand pain reported in Australia in the 1980s. While work-related arm and hand pain was and still is common, this particular epidemic was unusual in that it involved workers not previously considered as being at risk. At that time, the prevailing view was that these types of injuries were exclusively the result of exposure to physical hazards. However, the RSI epidemic primarily affected female keyboard operators. The condition itself was also characterized by reports of severe pain, numbness, and sensitivity to touch, while physical examination signs were often absent. The epidemic coincided with the introduction of video display terminals and keyboards similar to those used today. These changes decreased the need for regular breaks or postural changes (there was no need to change typewriter paper) and, combined with the new keyboards, allowed workers to increase their keystroke rate.

rapidly into an indispensable office tool that would profoundly change how work is done, one that deserves specific workplace layout guidelines as well as revised work procedures and new management attitudes.

Who would have thought, around 1990, that the computer would so profoundly change the job description of the typical office secretary: first into that of a word processing specialist, then of an administrative assistant or coordinator? Who would anticipate that every office staff member, manager, and even "the boss" (and all the professors) would do so much keyboarding themselves—work that, just a few years ago, would have been delegated to secretaries or assistants? Most of us also did not foresee that much of our "office work" is now done outside an office and on a laptop, tablet, or phone in a home office, a corner café, and even on a lounge chair poolside.

HANDHELD AND LAPTOP VERSUS DESKTOP COMPUTERS

The laptop computer is the direct high-tech descendant of the old typewriter. The laptop has a small keyboard with not many more keys than the typewriter. The display sits about where the platen of the typewriter used to be. The screen is placed just right for human vision, close to the keys (which are another important visual target) so that we look down at them all. (See Chapter 8 for more on this topic.)

The laptop's small size makes it possible to put it in virtually any location convenient to us at the moment. We can use it exactly as the name implies—resting on our lap—or of course position it on a table, desk, or any other feasible keyboard support. It can be placed anywhere in a sit-down or a stand-up workstation. We can take it wherever we want to go. As a result, the laptop computer allows us to change location and posture at a whim, which has health benefits of its own (see Chapter 5). Laptops provide much more mobility and versatility than the bigger and heavier desktop computers. And most users quickly adjust to the smaller size and different keyboard. Of course, a larger display and an additional keypad as well as other peripherals can be connected to the laptop when it is set up in the stationary office if the user prefers.

For traveling and easy use outside the office, handhelds and laptops are terrific tools. Taking it a step further, using a well-designed portable computer in the conventional office is perfectly acceptable and increasingly popular—and, once again, makes moving around rather than sitting still easier.

With a bit of imagination, we can even see ourselves in a future work environment in which we are free to move around, standing or even walking, because we will use our voices or gestures to communicate with the system. Or perhaps, if keyboards are still needed, they will be wireless (which we already have today), or even virtual, floating in space so we can place them wherever we want. Consider how much our work lives have changed even just in the past decade; our day-to-day lives are imminently more streamlined, fast, and connected. In the early days of the new millennium, there was no social media per se, nor did we have the cloud. You would lose all of your data if you experienced a dreaded hard drive crash, where now we easily back up our work via the cloud or via a miniscule flash drive.

Current times find us incredibly fond of our handheld "palm-sized" computers. Yes, we live in an era in which cell phone addiction—nomophobia—is a "thing" (see the story below). In the beginning, a handheld's purpose was less for writing extensive documents and more for brief notes and, naturally, for communicating with others. Now, the handheld computer is replacing the notebook and calendar of yesteryear for many of us and is even taking over many functions of full-fledged computers. In fact, the handheld is substituting for—or replacing—plenty of other devices as well. We stream shows and movies on our handhelds—does anyone remember trips to the video store ("be kind, rewind")? We take photos on them, we let them navigate for us to find our way around town, and we use them to replace our laptop when we are on the go.

Marie is a young Parisian whose active social life augments her busy career. She meets up with friends every Thursday night for dinner at a restaurant they chose earlier in the week. Her friends have noticed that Marie is growing increasingly attached to her cell phone—she texts throughout the meal, monitors her work e-mails over cocktails (even though she is officially off the clock), "pins" the coffee table she covets for her apartment as the group's appetizers arrive, and compulsively checks her Twitter feed to see if her last few tweets about her entrée have been favorited yet. When a lively debate arises about enacting tolls to ease city traffic congestion, she turns to Google Search for prevailing opinions and then searches Instagram to see what her friends elsewhere are ordering for dessert. Her friends are annoyed, and maybe they should be concerned as well. It turns out that smartphone addiction is not a joke; experts caution that it is a tangible and growing problem. In fact, nomophobia, the fear of being without one's mobile device, is seen as a significant enough issue to warrant treatment.

With their burgeoning popularity and impressive features, palm computers on the market are ever evolving. When we wrote our first edition of this book (in the year 2000), handhelds were nifty but lacking. Displays were small and tough to decipher, buttons were difficult to operate, and styli, if attached, were awkward and clumsy to use. For a while, handhelds were growing smaller; now, quite a few are reversing that trajectory and are growing larger. Further developments, we hope, will rely heavily on ergonomic research and recommendations; better human engineering will help propel the handheld computer as it forges even more inroads in offices and households.

HOME OFFICE

If you are of a certain age, you will remember that in an office environment, our ambition was once to attain the executive suite, the big corner office, the one with the private rest room and the bank of secretaries, and the thick cushiony carpeting. Then came the advent of the home office, where the goal became carving out a separate room with an intricately designed work space and a cutting-edge computer, although the reality was often a corner in the den, a spot in the kitchen, or, at best, a spare room, rustled up to do occasional paperwork for an hour or so.

Now that working remotely is commonplace and our devices are increasingly wireless and imminently moveable, the home office, as it were, is radically different than it was just a few years ago. The cloud has replaced file cabinets, Wi-Fi is ubiquitous, and power cords are disappearing at the same time as device charging options are multiplying. These developments leave remote workers with a dizzying array of places to work, as spaces formerly designated for work morph into work-life areas.

These realities both ease and complicate resulting ergonomic considerations. There is no single design solution for the "typical" remote worker, because there is no typical remote worker. Some of us prefer sitting at a desk or a kitchen table, while others use a laptop and handheld while lounging across a couch or standing at a lectern. Others—this author included—shake it up, alternating from one work area to another, alighting at a kitchen counter, then a kitchen table, then a couch, then a desk, then switching it up again. Consequently, we cannot offer hard-and-fast set of ergonomic rules for a conceptualized home office. Instead, we offer some tips and guidelines that take into account the type of work, the amount of space, the need for privacy and solitude, and the worker's physical and emotional characteristics:

- *Understand your work needs*: Do you like complete quiet or can you tune out background noise? Are you able to work among distractions or do interruptions derail your ability to complete your work? Answering these questions will guide your selection on whether or not you need a space that you can close off for privacy. Additionally, understand your preference for routine—some people prefer to hop out of bed and start working right away, while others might exercise or socialize before they ease into work.
- *Understand your physical/comfort needs*: If you are the kind of person who likes to move around and leave the home office for stretches at a time, check out the now-widely-available apps that allow you to identify sites where you can find Wi-Fi and some refreshments. Some workers like to meet with others and expand their professional network; plenty of remote workers set up informal gatherings at specified intervals (i.e., biweekly, monthly). These individuals might also want to check out shared office spaces; we cover these a bit further in Appendix B.
- *Invest in top quality furniture*: By "top quality," we do not mean "most expensive"; rather, you select furniture that fits you. Furniture should adjust to fit your own size and preferred working habits and should never force you into uncomfortable and even harmful positions—think slouching, leaning, and slumping over. Make sure that you have ample room to stretch your legs and adjust your position. And, of course, you should have a chair that really fits you and that integrates well with the rest of the office furniture.
- *Streamline and camouflage where you can*: Since this workspace is also your home, consider furniture that can also conceal work-related materials

and supplies. We are heading in the direction of wireless environments, but at this writing, some cords and wires are still necessary; make sure they are concealed and secured as well. This is also a safety issue, of course; we cover that topic more extensively in Chapter 7.

Bottom line: The home office deserves your full ergonomic attention because it can and should be "personalized" to suit you, in the interest of both your well-being and your performance.

CHILDREN'S COMPUTER WORKSTATIONS

The computer has not only changed the working conditions for many adults but is now a popular toy and learning tool for kids at home and an essential instructional instrument in the schools, both elementary and secondary. When the original issue of this book was published, ergonomists worried about the fact that young children were using workstations originally designed for adults, with seats and tables that did not fit, mostly because the furniture was too big (Hedge et al., 2000; Saito et al., 2000; Straker et al., 2000). Think of children craning their necks to gaze up at a computer display all while sitting in a too-big chair, feet not even meeting the floor. A standard computer keyboard, with its large size and excessive number of keys, also did not fit the small child's hands. Fortunately, today, computer workstations that fit children and can be adjusted to their quickly changing dimensions and habits are widely available. Also, just as laptops and handhelds are replacing desktops in the office, they are doing so in homes with kids as well. In using laptops and tablets at home, children can place them in comfortable positions for themselves.

At school, tablets and, increasingly, chromebooks are widely used and proliferating quickly. The chromebook, a browser-based hybrid between a notebook and a netbook, is becoming especially pervasive. It is inexpensive, easy to use, and requires little maintenance. It boots up quickly and is malware and virus resistant. The battery life of this device is extensive and it provides storage on the cloud, which translates into less file management and back-end administrative work for staff (Herold, 2012). Many educators particularly laud tablets and chromebooks because they encourage collaboration between teachers and students. For example, a group of kids can simultaneously edit a single document, helping each other and adding content in real time. We expect that tablets and handhelds will continue forging inroads, particularly since common core and other online assessments are here to stay and school budgets remain slim.

Given these realities, the same guidelines we offered earlier—understand the work needs, understand the physical/comfort needs, invest in top quality furniture, and streamline and camouflage where you can—should be used to make sure children are playing, learning, and working comfortably and safely.

OFFICE TASKS

Offices differ wildly in terms of size, location, layout, organization, and purpose, and there are even more different ways to categorize the work that is being done in them. Since this book is concerned with the ergonomic aspects of task performance and resulting workload, we divide office tasks into four gross divisions of major work categories:

1. Preparing texts, social media posts, and correspondence, such as letters and invoices, by computer
2. Keeping track of and recording payments, schedules, and events
3. Filing and retrieving material by hand or electronically
4. Talking and interacting with others by phone or face to face or screen to screen

The first two tasks, preparing texts and correspondence and keeping records, carry with them the risk of musculoskeletal overexertion injuries; per unit of working time, the number of key activations is even larger on computer keyboards than it was with the mechanical typewriters that once populated offices. With keyboards, the number of digit motions exceeds the capabilities of many computer operators' hand tendons, and, as mentioned earlier, many of us other than the employees formerly designated as secretaries currently do keyboarding on a daily or routine basis.

Filing and retrieving material electronically also makes the office worker susceptible to repetitive motion injuries, although presumably the number of digit motions per unit of working time would be lower than in the first two tasks. Filing documents manually and retrieving material is actually beneficial as it promotes moving around, especially if the filing cabinets are located away from the workstation. (An interesting irony for the industrial engineer is realizing that *not* having all work tools within close reach has its advantages.)

Only talking with others has relatively little dependence on the use or nonuse of computers; of course, if a telephone is the tool that is used for communicating, cradling a phone in between neck and shoulder over extended periods of time becomes an ergonomic issue as well. Fortunately, headsets are designed for just those purposes.

WORKING HOURS AND "POWER NAPS"

Concurrent with the spreading use of computers, handhelds, Wi-Fi, and the resulting constant connectivity is the reality of extended working hours per day. Our daily work schedules have expanded in recent years, at least in the United States: the proverbial "9 to 5" workday in the office, 5 days a week, is no longer the limit for many people. More working hours per day, in addition to work on weekends, place a heavy physical and mental load on a person and shorten the time available for recuperation, relaxation, and doing what we like to do.

We add an aside here: although our work hours are long, it is also true that we waste increasing time at work while on the company clock. Constant connectivity and ubiquitous wireless connection causes us to work more at home—and waste more time at work. Consider the time you might spend surfing the web, shopping online, exploring social networks, retweeting a friend's tweet, and checking personal e-mail—of all workplace distractions, the Internet is a huge productivity drain.

There is no hard limit for hours spent working that applies to every task and every person. Prevailing literature supports the hypothesis that 8 hours during daylight is an appropriate work duration for most office tasks. We know from our own experience that such 8-hour shifts can be almost insufferably long if the workload is high and continuous (or, for that matter, boring and monotonous) or that we can perform well for a longer work shift when the load is reasonable, stimulating, and interesting. Most of us are especially productive or motivated when we can set our own time schedule. Some people, like journalists or poets, habitually work during unusual times of the day and for long stretches of time. Yet, for us regular office workers, 8 hours tend to be enough, and many of us do best during the morning while some people are highly productive during the afternoon. We cover more information concerning length of work days and weeks in Chapter 2; the discussion on shift work or nontraditional work hours is covered in Chapter 7.

For most, a short break, possibly even a brief nap, after lunch is important. The literature acknowledges the existence of the after-lunch dip but cannot pinpoint one certain definitive chronobiological reason for it (an example of missing scientific explanation for a common phenomenon). For some of us, the need for the postlunch nap is related to the desire for peaceful digestion of food and drink, together—quite reasonably— with the habit of recharging our batteries for more work. Many of us are forced to conceal taking our "forty winks" although napping openly (as reportedly done, for instance, in Japan in many companies) would be beneficial. Interestingly, some companies appear to be acknowledging the benefits of a nap, making nap rooms available to employees. A large advertising agency in Chicago, for example, installed nap rooms in its flagship office building in the late 1980s, letting employees sign up for the rooms in half-hour increments. (Of course, although some companies do sanction nap times, getting your boss to approve of the daily snooze is a different matter.) As an aside, the same company also built and outfitted a corporate gym for those employees for whom a lunchtime run or workout was more energizing than a nap. Nonetheless, Winston Churchill, Napoleon Bonaparte, Albert Einstein, and some U.S. presidents (Kennedy, Reagan, and Clinton) habitually took naps (Brody, 2000).

There are convincing physiological and psychological reasons for taking short breaks often, in addition to a longer pause about halfway during the work shift. Some experts suggest taking a prescriptive approach to breaks, noting that people report feeling more energized and better able to concentrate after breaks. Both older and recent research has shown, again and again, that frequent short interruptions of physical work restore the ability to continue working at a high level. (Sustained keyboarding is tough on musculoskeletal components of the hand and forearm, and occasional breaks help reduce stress.) Taking a pause of 5 minutes every half hour is better than 10 minutes once an hour even though the total break time is the same (Balci et al., 1998; Kroemer et al., 1997, 2001; Neuffer et al., 1997).

GOOD OFFICE ERGONOMICS MAKES FOR IMPROVED ECONOMICS

Although the research, completed in 1999, is somewhat dated, the point is still valid: Liberty Mutual Insurance Company reported on a test of the theory that by giving employees more control over their environment and a better understanding of ergonomic principles, their performance would improve and their health problems diminish. The results of the 18-month study confirmed the expected: combined with ergonomic training, the flexible workspace increased individual performance and group collaboration. This was accompanied by a nearly one-third reduction in back pain and a two-thirds reduction in upper limb pain among the employees who had more control over their environment.

Every year, hundreds of thousands of employees miss work because of injury on the job; obvious causes of injuries like improperly lifting heavy objects may come to mind first, but some of the most insidious and costly injuries happen in sedentary work. Sitting at a computer and keyboarding for long periods of time can be extremely damaging from a repetitive stress injury (RSI) perspective. And with RSIs, not only does the employee suffer discomfort, pain, and possible short- and long-term injury, but the employer is on the hook for medical bills, lost work time, and temporary—or even permanent—disability.

Taking a proactive approach and helping employees find a comfortable, efficient way to do their work makes sense in both the short and the long terms. In addition, knowing that their opinion matters and their comfort is paramount shows employees that their company cares about their needs, and this might well improve employees' morale and productivity in the short term. In the long term, pain that could have become a chronic issue is avoided. In sum, implementing sound ergonomic policies and practices will prevent employee injuries, save money in the long term, and make for healthier and happier employees.

ONE DESIGN DOES NOT FIT EVERYBODY

This text contains specific design recommendations, for example, in chair dimensions, which have been derived to fit mostly North American and European users. Their body sizes are fairly well known, while those of other populations on earth vary. People with dramatically different body sizes would need their own specific ranges in furniture dimensions and adjustments. In addition to differing body dimensions and measurements, working postures and habits also vary substantially among people who inhabit the world—we like to sit in chairs, but our counterparts across the globe may prefer to kneel, squat, or recline. Finally, national regulations and standards also vary, sometimes dramatically.

Humans live in diverse social and organizational structures; have their special ways, preferences, and dislikes; and are accustomed to their own climates, sounds, and colors. Please use this text accordingly; recommendations given here will probably need to be modified to suit users in parts of the globe other than North America and Europe.

OFFICE DESIGN COMBINES SCIENCE, TECHNOLOGY, AND ART

Office design has evolved dramatically over the past centuries, and you can read much more about that in Chapter 4 or in Appendix B of this book. Office design has ranged from the small, dank offices of the 1500s through large, cramped workrooms that prevailed during Taylorism, followed by the tough concept of modular spaces and cubicles, with more luxurious private offices for those lofty souls ranking higher in the management hierarchy. Today's offices have evolved yet again, and tomorrow's version will reflect the prevailing technology, science, and art of the future.

Approaches to office designs are intuitive but also reflect philosophical and social ideology of the era in question. They are strongly based on the types of technologies that are available at that given time, such as heating and cooling systems, and on the predominant work methods: from writing by hand to typewriting to computer use and from telegraph to pneumatic tubes to telephone to wireless electronics. Technology is always in flux—what is current today may well be thoroughly outdated tomorrow. New technologies can be forgotten or become common almost overnight, as the telegram and e-mail demonstrate.

Technology and science also have influential roles in the design of equipment in the office. It is not enough to design an office based on albeit necessary science-based biomechanical knowledge since so many factors are at play in making an office comfortable and aesthetically pleasing. Consider that lighting engineering alone cannot make an office appealing to the eye; air-conditioning per se does not make all office workers comfortable. Ergonomics incorporates the scientific and engineering disciplines that are concerned with the human at work, but it does not stop there; the ergonomist still must work with the artistic, intuitive, daring, and speculative designer of furniture and office spaces to create that ultimate office space.

Science influences office design by providing theories about efficient and humanistic work systems and models. Science also provides methods and techniques to measure the outcomes: subjectively, such as by satisfaction or motivation, and objectively, such as by productivity or cost-effectiveness.

The literature also shows that, in the past, management determined—primarily according to economic considerations—which one of the many possible office designs would be selected. This probably was, and is, a reasonable process, but the assessment of what is good for the individual working in that office is (or should be) an important component of the decision making. Still, there are many irrational, intuitive, wishful, and unsupported ideas that have influenced judgment, such as "we think faster on our feet than on our seat," "posture affects mood and willingness to work," and "wall colors with red in them make people excited and aggressive." This should in no way compel us to round up and discard all of our office chairs or to paint the walls gray-green.

Doubtlessly, there are visionary and speculative traits in office layout, and without them some daring new designs and insights would not exist. Yet, people are the producers of office work, they derive new ideas and procedures, and they determine success or failure; they are the drivers and pilots, and—as with the cockpit in cars and planes—their office must be designed around them. Ergonomics supplies the information needed for that human-centered design.

REFERENCES

Balci, R., Aghazadeh, F., and Waly, S. M. (1998). Work-rest schedules for data entry operators. In S. Kumar (Ed.). *Advances in Occupational Ergonomics and Safety*. Amsterdam, the Netherlands: IOS Press, pp. 155–158.

Brody, J. (2000). Personal health: New respect for the nap. *New York Times*. http://www.nytimes.com/2000/01/04/health/personal-health-new-respect-for-the-nap-a-pause-that-refreshes.html. Accessed March 10, 2016.

Hedge, A., Barrero, M., and Maxwell, L. (2000). Ergonomic issues for classroom computing. In *Proceedings of the XIVth Triennial Congress of the International Ergonomics Association and 44th Annual Meeting of the Human Factors and Ergonomics Society* (pp. 6-296–6-299). Santa Monica, CA: Human Factors and Ergonomics Society.

Herold, B. (November 2012). Chromebooks gaining popularity in school districts. *Education Weekly* 34(12), 10–12, http://www.edweek.org/ew/articles/2014/11/12/12chromebooks. h34.html.

Kroemer, K. H. E., Kroemer, H. B., and Kroemer-Elbert, K. E. (2001). *Ergonomics: How to Design for Ease and Efficiency* (2nd ed.). Upper Saddle River, NJ: Prentice-Hall.

Kroemer, K. H. E., Kroemer, H. J., and Kroemer-Elbert, K. E. (1997). *Engineering Physiology* (3rd ed.). New York: Van Nostrand Reinhold—Wiley. (4th ed., 2010, Springer.)

Neuffer, M. B., Schulze, L. J. H., and Chen, J. (1997). Body part discomfort reported by legal secretaries and word processors before and after implementation of mandatory typing breaks. In *Proceedings of the Human Factors and Ergonomics Society 41st Annual Meeting.* Santa Monica, CA: Human Factors and Ergonomics Society, pp. 624–628.

Saito, S., Sotoyama, M., Jonai, H., Akutsu, M., Yatani, M., and Marumoto, T. (2000). Research activities on the ergonomics of computers in schools in Japan. In *Proceedings of the XIVth Triennial Congress of the International Ergonomics Association and 44th Annual Meeting of the Human Factors and Ergonomics Society.* Santa Monica, CA: Human Factors and Ergonomics Society, pp. 7–658.

Sauter, S., Dainoff, M., and Smith, M. (Eds.) (1990.) *Promoting Health and Productivity in the Computerized Office: Models of Successful Interventions.* London: Taylor & Francis.

Straker, L., Harris, C., and Zandvliet, D. (2000). Scarring a generation of school children through poor introduction to technology in schools. In *Proceedings of the XIVth Triennial Congress of the International Ergonomics Association and 44th Annual Meeting of the Human Factors and Ergonomics Society.* Santa Monica, CA: Human Factors and Ergonomics Society, pp. 6-300–6-303.

2 Working Well with Others

The most important single ingredient in the formula of success is the knack of getting along with people.

—Theodore Roosevelt, 26th President of the United States

What I don't like about office parties is looking for a job the next day.

—Phyllis Diller, comedienne and actress

OVERVIEW

This chapter deals with the primary macroergonomic issues that largely determine the role of every person in the system:

- How the individual employees feel about their company—is this "a good place to work"?
- How much they want to be involved—"I don't know, I just work here."
- How important they consider their work—"this place is nothing without me."
- How people get along with each other at work—"my colleagues are like family."
- How much effort they put into their work—"good enough for government work."

ORGANIZATIONAL BEHAVIOR

The field of study that helps us understand and deal with the interpersonal and organizational challenges in the workplace is called organizational behavior. While knowledge of many of the quantitative "microergonomic" aspects which we cover in Chapters 4 through 10 is essential, we also need to be sensitive to the qualitative aspects of our organization. We will briefly explain in the next paragraph why the "touchy/feely" characteristics of the office have a tangible ergonomic impact on employees.

Today, many companies are trying to keep their employees "happy"—satisfied and motivated. This is not sheer altruism at work; instead, organizations are simply recognizing that improving employee satisfaction will improve bottom-line profit. Happy employees are productive; they treat customers better, work harder, and even take fewer sick days than disaffected, disenchanted employees.

Moreover, happy employees tend to stay with the organization, which reduces one of the most significant costs to employers—those related to employee turnover. Not only does employee exodus carry a huge expense in terms of recruiting and training new employees, but a new hire accomplishes substantially less initially—in some studies, more than one-third less in the first 3 months—than an experienced worker. The outright expense of replacing a valuable employee can range from half to several times a year's pay. Finally, there is the indirect cost of unhappy employees who stay on the job: not only should we consider their low productivity and poor customer service, but we must also include the stress (and, potentially, serious stress-related illnesses) that they suffer as they trudge through what they consider a dreary work routine.

ELEMENTS OF AN ORGANIZATION

To understand an individual in her or his job, we need to comprehend not just the person but the environment—the organization—in which the person operates. To understand an organization, we must consider all of its components; organizations are networks of related parts, and all of the elements work together to enable the organization to function as a whole. This chapter will cover macro- and microviews of the organization, starting with the elements that define organizations and ending with the people that work there.

The basic organizational model, shown in Figure 2.1, depicts the internal and external elements that define an organization. Six structural elements interact to define a company: *strategy* (the company's plan for success), *structure* (corporate hierarchies), *policies and procedures* (company rules and regulations), *systems*

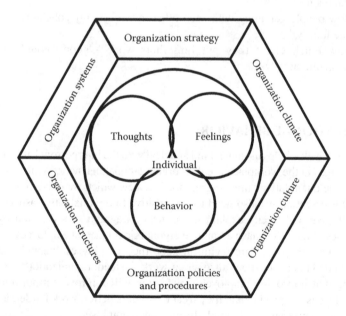

FIGURE 2.1   Basic organizational model.

(methods of allocating, controlling, and tracking corporate resources), *climate* (employees' feelings about the company), and *culture* (behaviors and feelings within the company). Each is explained further in this chapter.

THE INDIVIDUAL IS IN THE CENTER

As shown in Figure 2.1, the individual is at the center of the organization's operations, because ultimately, companies are made not of theories, structures, and machines but instead of living, breathing—and interacting—people. People affect the organization and the way it functions and, in turn, the organization affects these individuals. Why people act the way they do—and what we can do to keep them satisfied and successfully interacting with each other at work—is a fascinating puzzle that has kept psychologists and behaviorists busy for many decades.

BEHAVIOR

Many theories exist to explain an individual's behavior, including the "APCFB model"—the acronym stands for "assumptions," "perceptions," "conclusions," "feelings," and "behaviors." This model posits that a person's closely held assumptions color the perceptions of a given event, leading to highly individual conclusions; these in turn cause feelings that result in behaviors. Since everyone has a different set of assumptions, individual behaviors or reactions even to one common event can be widely divergent.

MOTIVATION

The attempt to understand the motivation of employees underlies behavioral theories: motivation is the attitude toward attaining a goal. The most widely cited original theories to explain behavior and motivation include Maslow's hierarchy of needs and Alderfer's existence, relatedness, and growth (ERG) theory (Figure 2.2). The basic premise of both theories is that individuals continuously strive to satisfy certain needs and that this quest in turn drives our behavior. We expound on both later in this chapter.

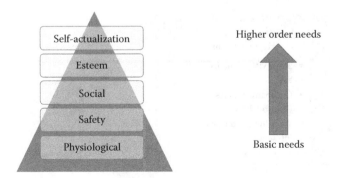

FIGURE 2.2   Maslow's and Alderfer's motivation theories.

Another well-known explanation is not internal or need-based but rather external; called reinforcement theory, at its core is operant conditioning, originally conceived by B.F. Skinner. Reinforcement theory's three components are stimulus, response, and reward; the stimulus invokes a response, which is subsequently rewarded (or punished).

Job Satisfaction

Job satisfaction is closely correlated to motivation and, as might be expected, several theories exist to explain the degree of pleasure an employee derives from his or her job. Some approaches postulate that job satisfaction is determined within the individual, when a person's physical and physiological needs are met; others center on the external factors of social comparison. Work conditions also may influence job satisfaction; Herzberg's two-factor theory is the most famous of the work condition theories. Herzberg believed that certain job content factors (like recognition and achievement) were positively associated with satisfaction. He postulated that other job conditions (like pay and the office environment) could generate dissatisfaction if perceived as negative (Figure 2.3).

Ultimately, all of these theories contribute to some extent to the explanation and understanding of motivation and satisfaction. All are covered in some detail later in this chapter.

Stress

Stress is a universal phenomenon; it can occur at all organizational levels and in all jobs. Stress refers to an individual's psychological and physiological responses

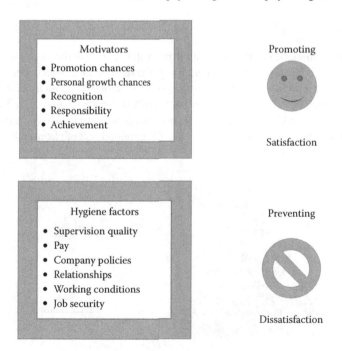

FIGURE 2.3   Herzberg's two-factor theory.

to environmental demands, and it sometimes carries severe consequences in terms of both physical and mental health. Reducing stress is critical for employee well-being and accordingly for the continued operational soundness of an organization. Techniques for coping with stress include setting and rearranging priorities, releasing emotions through actions or discussions, taking time off, moving on to new activities, or learning to accept a given situation. Some companies have implemented formal stress-reduction programs to help combat this invasive problem.

COMMUNICATIONS

Communication is key in any relationship between individuals; the relationships we forge at work are no exception. It is through communication that we establish hierarchies, foster motivation, exchange information, and express our feelings. One formal method of communication singled out for further examination in this chapter is the performance appraisal. In theory, an appraisal serves all the functions of communication; in practice, performance appraisals tend to be feared and avoided by both managers and employees. Fact is, they can be extremely valuable and even necessary when done properly. To make them most effective, appraisals should be done regularly and according to a set schedule, have tangible results directly tied to the outcome of the appraisal, and be conducted interactively, with both manager and employee participating in the discussions. Additionally, many companies utilize peer appraisals and upward (subordinate-to-boss) appraisals to glean colleagues' impressions of the individual and fully gauge his or her contribution to the organization.

We also cover workplace bullying in this section. Being harassed at work in this day and age is not acceptable, but it happens more often than one would like to believe. To combat this problem, which can render a workplace toxic, both employers and employees should have resources and plans to remedy the issue. We discuss some of those resources in this chapter.

QUALITY OF LIFE: AT WORK AND OFF WORK

Striking a balance between work and life outside of the job is a concern; a leanly staffed company in a tight labor market sets up a situation in which the employees might well be overloaded with work. Technological developments that connect us to work around the clock (such as pagers and cell phones and handheld computers) also influence the burden many employees feel. In our 2001 edition, we covered work–life balance in this chapter; now, in 2016, the topic has become even more important, and it merits its own chapter. Please read more about work–life balance in Chapter 3.

Note: You may skip the following part and go directly to the "Ergonomic Recommendations" section at the end of this chapter, or please read on for more detailed information on organizational behavior and how people interact in the office.

WHAT EXACTLY IS AN "ORGANIZATION"?

As depicted in Figure 2.1, the six main components of an organization are strategy, structures, policies and procedures, systems, climate, and culture. All interact with and affect one another, and as the environment changes, they too must adapt.

Strategy refers to the plan—stated or implicit—that the company has for success against its competitors. It guides the company's operations and determines specific tactics that the company will use to meet its overall strategy. Examples include the flooring company that offers next-day carpet installation services, the plumbing service that pledges to dispatch a plumber within 1 hour of a service call, and the grocery store chain that offers a far wider selection of organic and gluten-free items than competing stores.

Structures outline the hierarchies within an organization. In larger companies, they are usually depicted in organizational charts, which are detailed structural diagrams—most of us have seen several versions at some point in our careers. An organization's structure determines accountability and authority within its ranks; essentially, it defines the "official" relationships that exist between employees. Each level in a structured organization has its own degree of authority and responsibility; as you move up in the hierarchy, you increase your level of authority and responsibility. In general, each employee should only be accountable to one boss; this is known as "unity of command" principle.

Structures traditionally fall into seven main categories; these include function, product, customer, geography, division, matrix, and one called amorphous, if none of the aforementioned applies or works.

- *Function* means that work is divided by specific task, like finance, accounting, marketing, etc., and the various departments carved out by the structure generally report to an executive vice president.
- *Product* means that the staff is divided into all functions necessary to produce and sell a given product or brand; this structure usually features product or brand managers who are entrepreneurs in running their division almost like a separate company. Many consumer packaged-goods companies are organized in this manner.
- The *customer* structure divides all functions by customer need; this is the type of structure often found in service industries.
- *Geographic* structures organize work by location, and often, regional offices are set up to manage the business in a specific geographic area.
- A *division* structure means that each business unit operates like an independent organization under the overall umbrella of a parent corporation. Although a division may run itself almost autonomously, financing is often handled by the parent company.
- The *matrix* structure is unique in that an employee may have more than one boss because two or more lines of authority exist—and consequently, this structure does not adhere to the unity of command mentioned earlier. The matrix structure is generally reserved for organizations that feature complex and time-consuming projects that call for specialized skills. A management

consulting firm might have a matrix structure; another example might be found in a law firm, say, when a paralegal reports to several lawyers.

* *Amorphous* means that the organization has no formal structure; instead, employees forge and dismantle reporting relationships as needed. In reality, most larger companies use a mix of operational structures; accordingly, their structure is hybrid.

In the United States, employees within any of these structures are generally distinguished as line or staff. Line employees are those who are directly involved in producing the organization's products or services or meeting its primary goals, while persons who support these functions (through administration, advice, or general support) are referred to as staff employees.

Policies and procedures are the rules and guidelines that govern a firm's conduct. Policies are official rules and, in larger companies, are often formally written up in detail in an employee handbook. Examples of policies include the amount of medical benefits a company provides, guidelines and restrictions for posting on social media, or the number of paid vacation days an employee receives. Procedures are often not documented but are nevertheless widely understood by employees and are generally applicable to routine tasks. An example of a procedure might be how often e-mails are checked every day or whether or not refreshments are provided free of charge in the employee lounge.

Systems are developed by companies so that money, people, and things (machines, equipment, supplies) are properly allocated, controlled, and tracked. Systems fall into several categories of distribution and management, including money (accounting, investment, and budgeting), object (inventory and production), people (human resources, employee appraisals), and future (strategic planning, business development, marketing planning).

Climate and culture are closely related. Climate refers to the emotional state of the people in an organization; how they individually feel about the company, their coworkers, and their jobs. (Note that engineers use the term "climate" in its physical sense, referring to temperature, humidity, and the like, and we cover *that* kind of climate in Chapter 10.) Culture is more of a group phenomenon; it relates to the behaviors, beliefs, values, customs, and ideas that an organization encompasses. (Of course, use of the term "culture" here has little to do with its historical understanding.)

Consider two advertising agencies: one, a large and highly traditional company headquartered in a metropolis, and the other, a small "boutique" agency in a smaller town. Here are some of the characteristics of each: The first agency is 70 years old, has many long-standing blue-chip clients, and is highly structured, with formal dress codes, office procedure manuals, and written policies for all conceivable situations. Internal memos follow a set format and are heavily scrutinized by several administrative levels before they are issued; annual retreats and parties are scheduled and planned many months in advance. The second agency is 10 years old, with only a handful of young partners who employ highly creative individuals in an entirely unstructured office. Shorts and baseball caps are standard attire, no handbooks exist, and decisions are made quickly and with virtually no bureaucracy.

Clearly, each of these companies—although both operate in the same industry—has very different cultures, with widely diverse values and behaviors and hence with different climates. Employees in a company acquire its culture when they first begin working there; they become socialized into the broader group by interacting with the different members. Culture can be communicated subtly to new employees or loudly, with no ambiguity.

Climate and culture have become more important than ever, with management increasingly realizing that these elements are major determinants of employees' overall happiness within an organization. In this age of casual work environments, telecommuting, job sharing, and the emerging emphasis on individuality, companies' cultures vary more widely than ever before and are more important to people than ever before.

Climate and culture can be valuable bargaining chips to attract and retain employees. Recall the example of the advertising agencies outlined earlier. Each company attracts a different type of employee, and prospective recruits should understand the organization's culture before signing on. Employees can choose company cultures where jeans are appropriate office attire, pets can accompany their owners to work, on-site massages are routinely provided, and pool tables and nap rooms are available for employee "time-outs." Both the potential employee and the potential employer are well advised to carefully scrutinize the perceived "fit" of an individual within an organization: if a person does not appear to match the corporate culture, and vice versa, then an employee/employer relationship is often not effective and not advisable.

THE INDIVIDUAL

This brings us to the inner portion and most important feature of our basic organizational model: the individual. When a company's stock market value is determined, "hard" assets like property, plant, and equipment generally make up one-third to one-half of its value. The remainder is made up of "soft" attributes that traditionally received far less attention from shareholders: patents, customer base, and employee satisfaction. As a result, there is more incentive than ever to examine the inner circle of the diagram in Figure 2.1 to assess employees' state-of-mind and find out what influences it. "Human resources" are a company's most valuable assets.

Individuals are unique. All of us are different; our upbringing, environment, experiences, and personalities all make us special (yes, mom was right). Getting along with people is fraught with difficulties even in the best of circumstances, so throwing groups of people together in a work environment and expecting everyone to get along is unrealistic at best. Add to that the tensions inherent in a job—stressful or monotonous duties, long hours, and imposed reporting relationships with people you may or may not like—and problems are virtually inevitable. The popular media reflects this reality; there are vast amounts of books and e-books and articles and essays that provide instructions on dealing with employees and bosses, comic strips skewer them, and many movies and shows cover the treacherous terrain of interpersonal relationships on the job.

To gain insight into why people act the way they do on the job, we now take a look at some of the better-known theories of behavior, motivation, and job satisfaction. Please note that this review is by no means comprehensive; instead, it provides a glimpse of some of the most well-known alternative views.

WHAT EXACTLY IS "BEHAVIOR"?

APCFB Model

Many textbooks refer us to some form of the APCFB model; this model offers an explanation of how an external event can cause an employee's behavior. The letters in the acronym stand for assumptions, perceptions, conclusions, feelings, and behaviors.

The model postulates that the assumptions a person holds are an intricate part of the person's overall makeup; these assumptions are closely held beliefs that we have about the way the world and the people in it is, or should be. To a large degree, many of the assumptions we hold dear were created or at least strongly influenced early in our lives by parents and peers. These assumptions make up our value system.

When an external event occurs, we all see it through our own "filters" that influence how we view or perceive the event—assumptions affect a person's perceptions of the event. Our filters include internal defense mechanisms that act to protect us from psychological damage. These filters often prevent us from having an accurate reading of external events and of other people. When an event occurs, our existing value system—and the filters we have subconsciously created—shapes our view of the event into a given perception, which may differ significantly from reality. Yet our perception leads us to draw conclusions, which in turn lead to feelings, and these feelings then cause our behavior. Behavior includes doing and/or saying something, and this behavior may at times seem wildly out of context or proportion to the actual event, depending on the assumptions (and filters) that we held to begin with.

Looking more closely at these assumptions, we can further classify them based on how deeply rooted they are. Expectations are most easily changed; beliefs are more deeply rooted, and values are closely held assumptions; values are so strongly felt that they may not be changed at all or only slightly modified over long periods of time.

To illustrate this model, let us take an example: An insecure worker with a limited marketing background is promoted to marketing manager and asked to supervise the creation of the company website. When a web designer from an outside agency arrives to present the new website, considered by others to be brilliant, the newly promoted manager feels threatened by the web designer's knowledge while feeling that she herself ought to be more skilled in web design. To her staff's surprise, she rejects the proposed website out of hand in spite of its merits. Finding fault with the proposed website allows her to avoid confronting her own perceived ineptitude in web design (her filters helped protect her self-esteem). In the meantime, the new manager's staff is left with severe concerns about its online presence.

To augment our example, if company management had assured the new manager that her lack of online knowledge was not a problem and that she would be provided with appropriate training, she may have relaxed her expectation that she should excel in the field. Instead, her defense mechanisms caused her to act in a way that appeared irrational or out of proportion to her staff and her web-design agency. Her resulting behavior caused an unfortunate consequence that may cost the company significant money in lost sales since what might have been a solid online site was halted.

Understanding the value systems of employees—and then tapping into them—can be extremely effective. Doing so is often called "empowerment." Let us say that a given employee treasures her creativity but is given little leeway to express this creativity at work. If her manager were to revise the job to allow her to express this creativity, the employee will most likely become far more productive and happy at work. This is called "goal congruence"—the employee's goals and her manager's goals, in this instance, become equivalent or congruent. Goal congruence among the employees of a department or an organization makes the group far more productive.

As another illustration, consider a financial analyst who works at a midsized company that boasts a well-developed intraoffice computer system. This analyst was once a happy, energetic, and productive employee—but his mood and energy of late have taken a dramatic downturn. His wife has just had their first baby, and he has a very strong system of family values. While she plans to take a 3-month maternity leave, he would like to spend more time at home with his expanded family. He has a very long commute to work and generally spends 10 or more hours a day working and driving to work. His manager sees the toll that leaving his baby and wife on a daily basis takes on his mood and productivity. Their goals are congruent; both want to find a way to ease the analyst's separation pains and both wish to restore his former cheerfulness and productivity. Together, the analyst and his manager work out a schedule whereby he can work out of his home two days a week by telecommuting. Moreover, on the three remaining weekdays, he can adopt a "flextime" schedule whereby he arrives at work at 7 a.m. and leaves by 3 p.m., avoiding much of the traffic that snarls and complicates his daily drive. Within just weeks of implementing these changes, the financial analyst's work and attitude have vastly improved.

WHAT EXACTLY IS "MOTIVATION"?

Motivation incites, directs, and maintains behavior toward goals. Motivation and job performance are, of course, related; a motivated person who desires to do well at work is willing to expend effort to do so. Performance is the product of motivation and ability. Performance is moderated by situational constraints at work, which are factors that can stymie or enhance performance; examples are the climate (both psychological and physical) and up-to-date equipment.

Motivating employees is an elusive yet, of course, highly desirable goal; understanding what motivates people can help us understand each other and can ultimately give us clues on how to make people as effective and as happy at work as possible. A number of theories that explain motivation exist, and a few of them are outlined later; they can be roughly divided into two categories. The first group focuses on the *individual* and his or her inherent traits; the second group places the *environment* at the forefront of motivation. Theories that focus on the individual include need-based theories and the expectancy theory. Environmentally based approaches, which assume that motivation is driven by external factors, include equity theory and reinforcement theory.

Behavior appears to be motivated by the urge to satisfy needs. These needs characterize and drive us, and we act in a continuous quest to satisfy them. "Needs" are defined here as requirements for survival and well-being and are not optional to an individual. Needs fall into two related categories: physical, those that are necessary to physiological survival and comfort, and psychological, required for a fully functioning consciousness.

Maslow (1943) developed the widely known *need hierarchy theory*; later researchers then applied his work to organizational behavioral uses. Maslow and his disciples viewed motivation as a function of meeting an employee's needs, ranging from the basic physiological necessities (food, shelter, etc.) to the higher-order ones (self-esteem, self-actualization). He believed that individuals act in a never-ending mission to satisfy these needs, covering first the basic needs and then working their way up the hierarchy. He felt that lower-order needs were fulfilled before higher-order needs became desirable—the basic needs take precedence over the higher needs. The hierarchy of needs he proposed has five levels: physiological needs like the need for food and water are at the basis, followed by safety needs that focus on economic and physical security. The third rung revolves around social needs for belongingness and love, while the next level focuses on esteem needs, including self-confidence, recognition, appreciation, and respect. The highest-order needs are self-actualization needs; it is at this level that an individual achieves full potential and capability. Figure 2.2 depicts the model.

Alderfer (1972) took another need-based approach called the *ERG theory*. The acronym stands for three types of needs: existence, relatedness, and growth. This theory compresses Maslow's five needs into three. *Existence* needs are material needs and include food, water, compensation, and working conditions; *relatedness* needs involve relationships and interactions with others, including family, friends, and colleagues; and *growth* needs revolve around the desire for personal development and advancement. So if you're living below poverty level in an area beset by famine, your most important need is food; if you are affluent and well taken care of vis-à-vis all physical needs but are unable to visit any friends or family, your priority is getting access to your loved ones.

The two theories have similarities, as shown in Figure 2.2. Alderfer identified similar needs as Maslow did but saw these needs arranged in a continuum as opposed to a hierarchy, which allows for movement back and forth among the needs. He also hypothesized that if a person became frustrated in a fruitless attempt to satisfy a higher need, he or she would regress toward fulfilling lower needs instead; he called this syndrome "frustration regression."

Another approach is offered by the *expectancy theory*, which originated in the 1930s but was not applied to work motivation until the late 1950s; it then became quite prominent in the field of motivation research in the 1960s. The expectancy theory is a conglomerate of several researchers' ideas about motivation; it provides a clean, logical method to understand employees' motivation on the job and allows us to assess individuals' expenditure of effort at work.

Expectancy theory assumes that a person's motivation (and resulting satisfaction) depends on the difference between what his or her environment offers and what he or she expects. The theory can be expressed in the form of an equation that outlines the factors underlying motivation. In its simplest form, it states that

Motivation is a function of
{expectation that work will lead to performance} times
{expectation that performance will lead to reward} times
{value of reward}.

Here are some of the key tenets of this theory: it is a cognitive approach in which each person is assumed to be rational and knowledgeable about his or her desired rewards. The "reward" component in the aforementioned equation is generally something an organization can provide for its employees, like pay, promotions, and formal recognition. The "reward" might be better described as an outcome because it can actually be something negative as well (e.g., getting fired). It is also important to note that the "value" component (also called valence) is highly individual and completely subjective; it refers to how the employee rates the anticipated reward. Finally, the "expectancy" component is crucial; people must connect how hard they try and how well they perform. To illustrate, a factory worker on an assembly line may have little incentive to increase her rate of production since the overall speed of the assembly line remains unchanged regardless of her efforts. On the other hand, a real estate salesman can conceivably earn more the more houses he handles, so he may be motivated to conduct as many showings to potential buyers as possible during a given workday.

Each of the mentioned equation's components can help explain some aspect of motivation. We will again take an example: If an actor in a play performs well (expectation that work leads to performance) and receives critical acclaim for his acting in the piece (expectation that performance leads to reward), we may assume that he should be motivated. However, if money is the reward he values more than good reviews, and if he does not receive adequate financial compensation for his success, he may lose his motivation in spite of the reward given (critical acclaim). The expectancy theory has received prevailing support in the realm of organizational behavior and is widely applied today.

The *reinforcement theory*—also called operant conditioning—originated with B.F. Skinner and his work with animals in the 1960s. His findings were later applied to organizational behavior and used to describe motivation. The basic reinforcement model involves three key components: stimulus, response, and reward. The stimulus elicits a response; the reward is what is given to reinforce a desired response. Applying reinforcement theory to the workplace can be effective in motivating employees, but picking the appropriate rewards can be challenging. Individuals respond in different ways to varying incentives, as seen through the expectancy theory described earlier. Consider the dilemma faced by George, the manager of an upscale clothing boutique.

George manages an upscale clothing boutique. Currently, his six salespeople are paid on a commission basis, where each sale is rewarded with pay for the salesperson making the sale. This is how George has run his store for the past 8 years, but now he worries that the competition among his salespeople is becoming unhealthy. He is considering revamping the existing commission-based pay system and implementing a straight-salary system, which, he hopes, would foster teamwork. At present, salespeople argue over who will work the most coveted and valuable (i.e., busiest) shifts; they also compete ferociously with each other on the sales floor to obtain customers. They often rush customers into purchases to make their commissions, and several customers have complained about the overly aggressive salespeople. Although George recognizes the benefits of a straight salary, he wonders about a potential drop-off in sales without the commission structure since the current system directly rewards sales. Three of the salespeople favor a salary system to reduce interpersonal competition and infighting; the other three prefer a commission system because they feel it maximizes their income potential. The first three consider teamwork and a friendlier work environment a better reward than monetary compensation; the latter three find the opposite to be true.

Obviously, the reward must be meaningful and valuable to the employee—and there are plenty of individual differences among employees, so any "universal" reward systems are inherently imperfect. This in fact is one of the primary limitations of the theory—although in principle it functions well, it tends to ignore individual differences in how rewards are valued.

The *goal-setting theory* posits that people set targets and then purposefully pursue them; it is based on the premise that conscious ideas underlie our actions and that what motivates us are the goals we have set (Locke and Latham, 1984). These goals both motivate us and direct our behavior; they help us decide how much effort to put into our work. To positively influence motivation and behavior on the job, however, employees must be aware of the goal, must know what actions are necessary to achieve it, and must recognize it as something desirable and attainable. Difficult goals may foster higher levels of commitment on the part of the employee; and the more specific the goal, the more focused the efforts of the individual to attain it.

It is important to note that the workforce is constantly changing and developing. We have seen changes in what people value in terms of motivating factors, and we will continue to see this evolve. What once motivated people has shifted in recent years. Pay and stability used to be considered the important motivators in a job; now, with an increasingly diverse, lifestyle-conscious workforce, there are many other nonfinancial ways to reward employees and keep them motivated and happy. Childcare, on-site health clubs, flexible work hours, virtual work shifts, and vacation time—these are just a few examples of the rewards that employees might seek.

Consider that an estimated 80 million young Americans (born between 1981 and 1997)—the so-called Millenials—will be populating workplaces in the next few years. As of this year (2015), Millenials take over as the largest living generation in America. By 2020, one-third of all adults and half of all workers will be Millenials (Willer, 2015). As a group, Millenials tend to be educated (34% hold at least a bachelor's degree), culturally diverse, and creative; they also tend to dislike bureaucracy and traditional hierarchies and will change jobs frequently. Millenials are accustomed to constant connectivity and are digitally fluent. To motivate them, future employers should tap into what these workers find important. For example, Millenials are likely to want to help people in need, so employers could offer paid time off to volunteer. A recent Parade Magazine study (2015) showed that more than 60% of Millenials say they would rather earn $40,000 per year at a job they love rather than $100,000 at one they hate. And what they seem to crave most is work flexibility so they can balance work/life demands.

SUMMARIZING WHAT WE KNOW ABOUT MOTIVATION

In spite of how compelling theories are, there is no clear-cut single correct answer to the question of what motivates people. All of them, in our opinion, help us understand what "makes people tick" and forces us to truly consider what people in the office want and need. Accordingly, bits of all of the theories apply to different people, under different circumstances. Motivation is both intrinsic and extrinsic—factors within us and external to us drive our behavior. What most likely occurs is that we consider our own needs and wants and either consciously or subconsciously determine our goals; then we act to increase the chances of obtaining what we want. We do know from the theories mentioned earlier that we will strive first and foremost to fulfill basic survival needs like securing food and shelter and that beyond this our needs are still very real but vary widely among individuals in terms of priority and strength. Other people influence us and shape our motivation and behavior because we are all to some degree social creatures. How hard we work is influenced by what our work will bring us: if we are rewarded in ways that are meaningful and valuable to us, and if we perceive a definite link between the strength of our efforts and performance, we will work strenuously to perform well and gain those desired rewards.

WHAT EXACTLY IS "JOB SATISFACTION"?

Employees spend vast amounts of their time on the job, so it is not far-fetched to assume that job characteristics will influence job satisfaction, which will then affect productivity and performance at work. Job satisfaction is important to us as a society, as managers, and as employees for several reasons. First, as a society, we believe in a high-quality work life as a goal; we feel that everyone has the right to fulfilling, rewarding jobs. Additionally, employers are increasingly interested in keeping employees happy because job satisfaction is associated with critical (and revenue-impacting) variables like turnover, absenteeism, and job performance. And employees who are happy on the job might well enjoy a higher quality of life overall, with fewer stress-related disorders and illnesses, than those who are not happy. There does appear to be a causal link between job stress and physical disorders; stress is covered in further detail later in this chapter.

Job satisfaction is defined as the extent to which a person derives pleasure from a job. It is distinct from "morale," which refers to the collective spirit and overall goodwill of the larger group in which the person functions. Job satisfaction has been extensively researched and, accordingly, there are several theories that strive to explain what causes or prevents it. These approaches can be broadly categorized into need-based or value-based theories, social comparison theories, and job content and context theories.

According to need- or value-based theories, job satisfaction is an attitude that is determined within the individual. The theories in this general category postulate that every person has physical and physiological needs that one strives to fill in order to obtain satisfaction. A job defined as "satisfying" here would meet physical needs (like food and shelter) via rewards including appropriate income and would meet psychological needs (like self-esteem and intellectual stimulation) through growth opportunities and professional recognition. (We already discussed the most well-known need-based theories earlier when we reviewed motivation.) Needs not only motivate us, but they provide satisfaction when they are fulfilled.

Value-based theories recognize that although needs may be universal, individuals assign different weights and priorities to them. One person might value personal growth and recognition at work, for example, but care little about monetary rewards; this person would be satisfied even in a low-paying job as long as he or she was recognized for their performance and given plenty of professional challenges. According to these theories, job satisfaction is present when a person's needs and values are met; if a person's needs or values change, however, dissatisfaction could result.

While the need-based theories explain a great deal about happiness at work, they neglect to take into account that people do not work in a social vacuum and that human nature compels us to compare ourselves with others. This brings us to the second group of theories, which center on social comparison. Here, researchers posit that people assess their own feelings of job satisfaction by observing others in similar positions, inferring their feelings about their jobs, and comparing themselves to the other people who work in similar capacities. Intuitively, it makes sense that social comparisons influence job satisfaction: our perception of others factors into our lives in infinite ways, and satisfaction at work is no exception.

Work conditions are the focus of the third group of theories that attempt to explain job satisfaction; Herzberg's two-factor theory is the best known and best researched of these. Here, job *content* factors, when present and positive, are associated with satisfaction, while job *context* factors, when negative, are associated with dissatisfaction. Herzberg and his colleagues isolated five *content* factors that they felt were "satisfiers": achievement, recognition, the work itself, responsibility, and advancement. He proposed that a job featuring plenty of these factors would create job satisfaction, and that their absence would result in a neutral or indifferent employee. Further, he speculated that a different set of factors influenced dissatisfaction; these are the *context* factors. *Context* (also called hygiene) factors include company policies and administration, salary, supervision and management, interpersonal relationships, and physical conditions at work. He proposed that a job without good context factors would make an employee dissatisfied, while the absence of context factors would lead to a neutral or indifferent employee. The upshot of Herzberg's theory is that a job should ideally be designed with plenty of rewarding content factors (to ensure satisfaction) and positive context factors (to avoid dissatisfaction). Of note is that the opposite of "job dissatisfaction" is not "job satisfaction" but rather "*no* job dissatisfaction," because content factors and context factors are *not* the same.

Consider the case of Francoise and Simone, two former schoolmates who have forged very different careers for themselves. Both are extremely satisfied with their jobs. Francoise is a lawyer, on the proverbial fast track at a large inner-city law firm. She treasures the recognition that she gets for working long hours and winning the cases assigned to her, and her impressive record of promotions and steady career advancements help satisfy her desire for positive feedback and approval. Her employer's posh downtown law offices boast large, expensively furnished window offices; all partners receive an extensive support staff, generous expense accounts, and lofty annual salaries. In spite of the 80-hour-plus workweeks, she feels that the "fame and fortune" of the job make it more than worthwhile. Her school chum Simone, on the other hand, works as an animal technician at a non-profit animal shelter in the same city. She works in a squat, simple building in which most of the space is devoted to operating rooms and kennels and the pervasive smell is of ammonia. Everyone—including the veterinarians—handles even routine tasks like answering phones, greeting visitors, and mopping up when needed. The vets and animal technicians are a close-knit, informal group. Simone's salary barely provides her with enough money to meet expenses, but she gleans rewards from colleagues' verbal praise and from the sight of recuperating animals (and the animals' eventual owners). She appreciates the freedom of wearing shorts to work and rarely works more than 32 hours per week. This leaves her plenty of time for family and friends, her own pets, and her hobbies. Although both women have chosen very separate careers, with what Herzberg would call widely varying job content and context factors, both consider themselves extremely satisfied at work.

Ultimately, all the theories described earlier contribute to an overall understanding of what keeps employees happy on the job. Each theory supplies some degree of explanation for what contributes to job satisfaction. An integration of existing theories would likely provide the best approach to understanding satisfaction. Clearly, we all strive to fill basic needs—shelter, food, and adequate clothing are mandatory for us and take precedence over higher-order needs. Once we have ensured the basics, we want to be intellectually challenged, recognized and praised, promoted, and rewarded. However, we all have different definitions of intellectual stimulation or of what constitutes a reward. Also, we place different weights on what is important. Moreover, we are doubtlessly influenced by what we see around us—are our colleagues earning more than we are for doing the same work? Does our friend who works from home in jeans and T-shirts seem to be having more fun, or is it the one who wears designer suits and works in an office with a large window? Do we like our peers at work? How supportive is our boss?

Summarizing What We Know about Job Satisfaction

As a summary statement of the various factors underlying motivation and job satisfaction, we might do best to take a closer look at a company with high marks in job satisfaction. Google has ranked as the top place to work for 6 years in a row, according to *Fortune*'s annual survey, and consequently averages two million job applications a year. And here is what you can expect if you land one of those coveted jobs at Google:

- Excellent salary, with one of the highest tech salaries out there
- Fascinating and challenging work, with the company mission statement "do cool things that matter"
- Brilliant colleagues
- Creative empowerment, with employees given 20% of paid company time to pursue their own personal projects
- Excellent communications throughout the company
- A work environment designed "to create the happiest, most productive work-place in the world" where you don't want to leave campus, and including free, fresh, and organic food served in various themed cafes and restaurants

As we discussed in defining motivation, earlier, there are several compelling theories to describe job satisfaction, but there is no single correct answer to the question of what satisfies people in their jobs. Instead, bits of all of the theories apply to different people, under different circumstances, at different times in their lives.

WHAT EXACTLY IS JOB DESIGN?

We can also understand job satisfaction and employee motivation by looking at the way a job is designed. In the early 1900s, Frederick Taylor radically changed jobs by introducing specialization. He suggested that jobs be broken down into small tasks that could then be standardized and divided among workers. This usually meant that a given worker did the same task repeatedly, resulting in extreme specialization—and, unfortunately, often extreme tedium. Although Taylor's job designs generally improved productivity itself, workers rebelled against the boredom, depersonalization, and routine of their repetitive jobs and began showing up late, not appearing for work at all, and suffering from stress. (Highly repetitive work can lead to injury, as discussed with respect to keyboarding in Chapter 6.) In the mid-1900s, in contrast to Tayloristic job simplification, "job enlargement," increasing tasks and variety, and "job enrichment," increasing workers' participation and control, grew popular.

Several motivation and job satisfaction theories that we mentioned earlier can help to explain how job design influences behavior. The job characteristics model (Hackman and Oldham, 1980) synthesized much of the existing research on job design and became one of the most accepted and examined explanations of job enrichment. The job-characteristics model proposes that any job can be described via five core dimensions. These are (1) *skill variety*, the number of different talents and activities that the job requires; (2) *task identity*, the extent of work from beginning to end that the job involves; (3) *task significance*, the job's impact on others. These three factors make the job meaningful. The fourth factor (4) *autonomy* describes the degree of independence in planning, controlling, and determining work procedures. This makes the person feel responsible. Finally, (5) *feedback* provides information about one's effectiveness, how the performance is evaluated and perceived.

All of these core job dimensions, when present, influence "critical psychological states" of the jobholder through the experience of meaningfulness of work, responsibility for work outcomes, and knowledge of results. According to the theory, high levels of the three critical psychological states lead to favorable work outcomes. These include high motivation, higher quality work, higher satisfaction, and lower absenteeism and turnover. Put more simply, the basic tenet of good job design is that employee happiness leads to quality of work life; one in which employees can fulfill their personal potentials.

WHAT EXACTLY IS POWER?

The word "empowerment" has already appeared in this chapter; it is a popular buzz-word in corporate literature. In the intricate political landscape of an organization, power affects everyone. Power is a motivator; people tend to crave, use, and occasionally abuse their power, so it is helpful to understand where it originates.

There are five types of power: coercive, reward, referent, legitimate, and expert.

Coercive power is based on fear; people who have this power are in the position to inflict some sort of punishment on others. At a restaurant, the shift manager might place a waiter who is on her bad side on a "slow" station with low-turnover tables where tips will be disappointing. An organization's coercive power includes its capacity to fire employees or dock pay.

Reward power is just the opposite of this; it is based on the expectation of receiving praise. A teacher might design an especially detailed lesson plan in the hopes of receiving praise from the principal; the principal has reward power. A company's reward power lies in its ability to provide incentives such as raises, promotions, and paid vacations.

Referent power exists when people admire a person regardless of his or her formal job title or status; generally, this person has particular charisma and an ability to attract and inspire followers. An example might be the administrative assistant who is widely known by company employees as the person who single-handedly keeps the office running. She knows where to find everything, what the social media passwords are, when to reorder supplies, and which company coffee machine yields the best espresso.

Legitimate power is due to the formal status held within the organization's hierarchy. A personal trainer at a gym has legitimate power over his clients—even if just for the duration of their workout session. A corporate CEO makes the final decisions in corporate meetings because she has the authority and even the duty to do so.

Finally, *expert* power comes from one's own skill or knowledge, regardless of formal position or job status. A low-level computer programmer may have extraordinary power in a company if he is the only one who knows how to edit the website, for example, or to keep the corporate Facebook page up to date. Another example is the accountant brought in to set up the bookkeeping system of a fledgling company; her perceived expertise in the accounting field lends her this type of power.

It is possible for a person to have several types of power in an organization—indeed, a person could have all five. Consider the case of a well-known and respected orthopedic surgeon in a busy private practice who hires a medical assistant to help. The surgeon holds legitimate, coercive, and reward power over the assistant—she is formally the assistant's supervisor, she can fire her at any point, praise her, and give her raises. Additionally, her expertise in the field of orthopedics gives her expert power over her assistant, and her much-admired professionalism and knowledge lends her referent power as well.

WHAT EXACTLY IS STRESS?

Stress is a psychological state caused by environmental conditions that lead to a person's specific psychological, behavioral, and physical reactions. Stress does not discriminate; almost every worker in any job faces stress, and it occurs at all job levels. Because it is initially a psychological phenomenon, stress is closely tied to motivation, behavior, and job satisfaction.

Stress at work is real, as the National Institute for Occupational Safety and Health (NIOSH) points out in a 2014 publication called "Stress...At Work."

- One-fourth of employees view their jobs as the number one stressor in their lives.—*Northwestern National Life*
- Three-fourths of employees believe the worker has more on-the-job stress than a generation ago.—*Princeton Survey Research Associates*
- Problems at work are more strongly associated with health complaints than are any other life stressor-more so than even financial problems or family problems.—*St. Paul Fire and Marine Insurance Co.*

Stress as a psychological phenomenon involves interactions between an individual and the demands of the environment. The concept of stress has its foundations in research by Walter Bradford Cannon, early in the last century, and especially by Hans Selye (1978). They described what we now simply call "stress" as "general adaptation syndrome" (GAS), a pattern of physiological reactions to the environment, or the body's reaction to adverse environmental stimulation (Conway and Smith, 1997). Selye characterized GAS as a mobilization of energy resources that proceeds along three stages: alarm, adaptation (resistance), and exhaustion. In the alarm stage, the body mobilizes its resources to meet the assault of a stressor; these bodily resources include increased blood pressure and heart rate, elevated muscle tension, higher levels of hormone production, and the release of energy. During the adaptation stage, the body works hard to maintain homeostasis (physiological balance). This effort to normalize its systems strains the body. In the third and last phase, the person's biological integrity is endangered if the body's systems are overworked in their efforts to adapt—the "normal" stress has become an overwhelming "distress." At some point, if overload is continued, the primary biological systems begin to fail, and serious physiological problems can occur. In short, Selye believed that response to a stressor led to heightened use of resources in order to either resist or adapt to the demand. This applies in a similar fashion to psychological loads and the person's subsequent responses.

When an individual is experiencing psychological, behavioral, and physiological effects of stress at work, actual changes in body chemistry occur, and these changes—which include heightened blood pressure, increased muscle tension, decreased immune system response, and hyperventilation—have been shown to lead to depression, headaches, and ulcers, to contribute to an increase in cardiovascular health problems, even to an increased risk of work-related musculoskeletal disorders (Carayon et al., 1999).

Selye hypothesized that stress followed a stimulus-response model, in which certain environmental demands (stressors) would invoke a related response. This early theory has been updated, however, to take into account our individual differences. In a given situation, individuals react differently to the same stressors, and any one person may even react differently at different times. A whole host of personal characteristics determine the physical and psychological effect that any given stressor will evoke in an individual. These include personality, current state of physical and mental health, skills and abilities, physical conditioning, prior experiences and learning, value systems, goals, and needs. In order to take these individual differences into account, a cognitive approach to understanding stress is now widely accepted. It refines the straight stimulus–response characteristic of the earlier research.

Given that stressors vary in terms of the severity of their impact on people, we simply generalize here and state that there are certain environmental demands that tend to exert the most significant effect on employees. Some of the biggest stressors are the following:

- Demanding jobs with high workloads, pressure to perform, a too-fast work pace
- Lack of control over the process and work and lack of autonomy
- Task difficulty, with duties perceived as overly complex, or with conflicting demands
- Overbearing responsibilities for others, lack of social support, isolation
- Monotony and underload on the job, with overly-routine, repetitive, and boring duties and little content variety
- Poor supervisory practices, including nonsupportive superiors or incompetent management
- Technological problems, like frequent computer malfunctions or equipment breakdowns

Smith and Conway (1997) specifically researched the effects of computerization on job stress. Reviewing a number of studies done by various researchers, they concluded that psychobiological mechanisms do exist that link psychological stress to heightened susceptibility to work-related musculoskeletal disorders (e.g., in keyboarders) because stress can affect our hormonal, circulatory, and respiratory systems. This in turn exacerbates the impact of physical risk factors, such as improper workstations and excessively repetitive movements, discussed in later chapters in this book. One can argue, therefore, that proper ergonomic design of work tools and equipment is particularly consequential in high-pressure jobs.

Reducing distress becomes an important goal for all levels of the organization, from the individual employee through supervisors to the company as a whole. The fundamental approach is to reduce the workload and other factors on the job that are unnecessarily stressful. Lower stress levels help employees enhance job satisfaction and maximize personal health, help supervisors keep their staff happy and healthy, and help companies achieve and maintain lower healthcare costs and better productivity. A number of researchers have examined approaches for managing stress, as reviewed by Muchinsky and Culberson (2016), who listed six major coping mechanisms. These were as follows:

1. Setting priorities—deciding what is really important and what is not
2. Utilizing home resources, like talking about feelings with family members or friends
3. Recovering and dealing with the problem by moving on to a new activity
4. Distracting oneself with an unrelated activity or by taking the day off
5. Passive toleration by giving up and accepting the situation
6. Releasing emotions by taking feelings out on others

Clearly, the latter two—particularly the last one—are not outcomes we would like to see. What is more, and unfortunately, quite a few people deal with stress by smoking, drinking, turning to drugs, and withdrawing from others.

In recognition of the many negative effects stress can exert on employees, some companies have created and implemented formal stress-reduction initiatives and programs. These include time management seminars, instruction of relaxation techniques, on-site wellness facilities that include exercise equipment and yoga classes, on-staff psychologists or counselors available to employees, biofeedback seminars, and flexible schedules.

WHAT EXACTLY IS COMMUNICATION?

One activity that pervades and influences—directly or indirectly—all of the topics we have discussed earlier is communication. It is the exchange of information between two (or more) persons and the inference of its meaning. Individuals understand and establish their roles in an organization through many modes of communication: how they provide and receive feedback, how they make decisions, and how they state and pursue goals. Functions of communication include establishing control, disseminating information, enhancing motivation, and expressing feelings. Muchinsky points out that communication is the lifeblood of an organization—and that failure to communicate is its Achilles heel. Communication occurs throughout the organization and the environment in which the company operates—interpersonally (among employees), intraorganizationally (among the groups and departments of the company), and interorganizationally (between organizations). If you have ever perused the job ad boards, you have noticed how many companies request "excellent oral and written communication skills" in their job specifications: the value of positive communication cannot be overstated.

INTERACTING WITH OTHERS

Surviving and, ideally, thriving in the intricate political structure of an organization virtually mandate getting along with people. The people with whom you most need to interact effectively are your subordinates and your supervisor. Managing the relationship upward is as important as managing the relationship downward. In order to "manage your boss," you must understand your boss—including his or her context, goals, the pressures he or she faces, strengths, weaknesses, and preferred work styles—and then you must understand the same things about yourself. Once you have done this, you can map out, develop, and maintain a relationship that fits both your needs and styles and meets both of your expectations. Ideally, this relationship should be based on dependability and honesty and selective use of your boss's time (use this time to keep him or her informed—bosses rarely appreciate surprises!)

The steps mentioned earlier may seem unrealistic—how can we truly understand the stated and unstated goals of our bosses?—yet even just trying to understand them will bring us closer to effectively dealing with colleagues. If you know, for example, that one of your supervisor's goals is to complete a marathon in October, and his training schedule includes midday training runs, you will not schedule meetings over lunch. If you realize that your boss is introverted and dislikes face-to-face interactions, keep her informed via e-mail. These basic principles can be similarly applied to dealing with subordinates—understand them, understand yourself, and base your relationship with them on these assessments.

One tool that will help you to interact with supervisors, subordinates, and colleagues alike is a technique called "active listening." Many times, when we assume that we are listening, we truly are not; instead, we might be silently disagreeing with the speaker (*he doesn't know what he's talking about*), formulating our response to the person talking (*well, you could have sent a text*) or thinking of something else and tuning out entirely (*I wonder what I should have for lunch*). Active listening helps circumvent this problem by forcing us to really hear what the other person is saying. This lets us get a valid perception of what the speaker is communicating—and gives the speaker the satisfaction of truly being heard. When you practice active listening, you pay close attention to what the person is saying. You respond to the information the speaker gives you without leading and without giving advice, and you identify the speaker's feelings along with the content of what they have said by absorbing the information and repeating portions back to him or her to verify the information. Importantly, an active listener gives control of the conversation to the other party.

WORKPLACE BULLYING

Bullies are not just a bane of schoolchildren; they unfortunately appear at the work place as well. According to a Workplace Bullying Institute 2014 U.S. survey, workplace bullying is defined as repeated, health-harming mistreatment of one or more persons (the targets) by one or more perpetrators. It is abusive conduct that may include the following:

- Threatening, humiliating, or intimidating
- Work interference—sabotage—that prevents work from getting done
- Verbal abuse and, in very rare instances, physical abuse

We do not expect bullying at work, so we may not even recognize it when we come across it. We can categorize workplace bullying in two ways: there is the overt kind in which a person or people engage in oafish (at best) or outright hostile or violent (at worst) behavior, and there is also the insidious kind, where small, repetitive acts add up over time. Huppke (2015) calls them the "hammers" and the "needles," going on to add that the hammers must of course be addressed, but letting the less obvious needles go can swiftly turn a workplace toxic. And the effects on the employee being bullied can be truly devastating. A British study found that one person in three leaves the job after being bullied (Hunter, 2015).

It can be hard to know if bullying is happening at the workplace, because there can be a "fine line" between managing or coworking and bullying. Comments that are objective and are intended to provide constructive feedback are not usually considered bullying but rather are intended to assist the employee with their work. Consider the following actions on the part of a manager or colleague:

- Expressing differences of opinion
- Offering constructive feedback, guidance, or advice about work-related issues
- Assigning work
- Taking reasonable disciplinary actions
- Managing a staff member's performance

These are *not* workplace bullying behaviors.

While bullying is a form of aggression, as mentioned earlier, the actions can be both obvious and subtle. The Canadian Centre for Occupational Health and Safety (CCOHS) offers a helpful list of possible bullying activities; while this list is in no way comprehensive, it does illustrate some of the ways bullying may happen in a workplace. We should also remember that bullying is usually considered to be a pattern of behavior where one or more incidents will help show that bullying is taking place.

Examples include the following:

- Spreading malicious rumors, gossip, or innuendo that is not true
- Excluding or isolating someone socially
- Intimidating a person
- Undermining or deliberately impeding a person's work
- Physically abusing or threatening abuse
- Removing areas of responsibilities without cause
- Constantly changing work guidelines
- Establishing impossible deadlines that will set up the individual to fail
- Withholding necessary information or purposefully giving the wrong information
- Making jokes that are obviously offensive by spoken word or e-mail
- Intruding on a person's privacy by pestering, spying, or stalking
- Assigning unreasonable duties or workloads that are unfavorable to one person (in a way that creates unnecessary pressure)

- Creating a feeling of uselessness—underwork
- Yelling or using profanity
- Criticizing a person persistently or constantly
- Belittling a person's opinions
- Giving unwarranted (or undeserved) punishment
- Blocking applications for training, leave, or promotion
- Tampering with a person's personal belongings or work equipment

If you are not sure an action or statement could be considered bullying, you can use the "reasonable person" test. Would most people consider the action unacceptable?

Another way to determine if you are a target of workplace bullying is to examine yourself and ask yourself if you notice symptoms including

- Feeling nauseated and anxious before the start of your workweek
- Obsessing about work while at home or at other nonwork activities
- Experiencing physical health effects, like loss of appetite or inability to sleep
- Experiencing psychosomatic symptoms, like headaches or stomachaches
- Feeling vulnerable at work
- Feeling shame about what's happening at work and not sharing what's happening with loved ones
- Using paid time off for "mental health breaks" from the misery
- Losing confidence
- Losing enthusiasm, motivation, and enjoyment in work that you once liked and took pride in

Companies should care about workplace bullying, if not for compassionate reasons—which frankly would be enough for us to care—then for the overall health and soundness of the organization itself. Workplace bullying leads to increased absenteeism, employee turnover, accident/incident occurrence, reduced customer service, decreased morale and productivity—all of which reduce profitability.

Bullying is a form of violence in the workplace, so it is prudent that companies devise a workplace policy against harassment in the same way that employers put together a fire emergency policy, or a social media policy. A comprehensive policy should cover a range of incidents and clearly convey management's commitment to preventing workplace harassment. The policy should spell out exactly what to do if an employee feels bullied and put channels in place for reporting any such behavior. Importantly, the process should be clearly outlined, with strong emphasis that it is confidential, and no reprisals will be made against reporting employees. The policy should also encourage everyone at the workplace to act toward others in a respectful and professional manner, explain what behaviors are considered bullying—perhaps by providing clear examples of such behaviors—and that such behaviors are unacceptable, and assure that all complaints will be treated seriously, promptly, and in confidence. Finally, employers must train supervisors and managers on how to deal with complaints and how to arrive at solutions. If necessary, an impartial third party should be available to help resolve any outstanding harassment issues.

Importantly, the policy should apply to everyone associated with the company—to management and staff, of course, but also to clients, independent contractors, boards of directors, and any other individuals who work with the company. If at all possible, consider offering a confidential employee assistance program so employees with personal problems seek outside assistance. Also, victims of office harassment should be offered supportive services.

Workplace bullying is not routine, acceptable, or deserved. If you feel that you are being bullied or subjected to harassment, consider the following:

- If you feel comfortable dealing directly with the harasser, firmly tell the person that the behavior is not acceptable and ask him or her to stop. If you are uncomfortable approaching the individual on your own, ask a supervisor or union member to accompany you in approaching the person.
- Keep a journey or record of daily bullying events. Make sure you record the date, time, and what happened in detail. Also note the names of any witnesses and the outcome of the event. Bear in mind that the character of the incidents is important but so is the frequency or pattern that the incidents establish.
- Retain copies or images of any written documentation, texts, e-mails, tweets, posts, or other messages that come from the person perpetrating the bullying and that represent the harassing behaviors.
- Report the harassment to the person identified in your workplace policy, or your supervisor or manager; if your concerns are minimized or ignored, proceed to the next level of management or, if necessary, to a governing body such as a board of directors.

No matter how tempting, we strongly recommend not retaliating. This will cause confusion at minimum and may dilute any action that would have been taken against the bully. Even worse, you may end up looking like the perpetrator. (Adapted from "Violence in the Workplace Prevention Guide," CCOHS.)

Performance Appraisals

Performance appraisals are formal means of communicating among managers and employees. When done properly and at the right time, performance appraisals can be very useful in achieving organizational/administrative goals, for providing feedback and evaluation, and for coaching and development. *Organizational or administrative goals* pertain to personnel actions like placement, promotions, and pay (and, when need be, provide the necessary documentation for firing). *Feedback and evaluation* refer to the employee's performance, including strengths and weaknesses of the employee's work. *Coaching and development* relate to the critical final goal of the appraisal: how to improve—rather than punish—the employee's work. Working together, the manager and the subordinate should agree on specific goals and timetables for improvement; this provides a valuable opportunity for encouragement and career coaching.

While they are very valuable in the theoretical sense, in practice, performance appraisals are often misused, mismanaged, and widely feared. People resist formal performance appraisals for a number of reasons. Managers may dislike giving evaluations because they are uncomfortable criticizing subordinates and feel that they lack the skills needed to evaluate employees. Subordinates often become defensive and anxious during their appraisals. Accordingly, the task is often delayed until the appraisal has effectively lost its usefulness. Management plays a vital role in making the appraisal process succeed by tying the process into the organization's overall role. Appraisals must have tangible results; for example, if promotions and raises are given based on performance, the appraisal process should be the tool for determining performance. Put differently, there should be a high correlation between the thoroughness of the appraisal process and how extensively the resulting information is used.

Performance-appraisal systems often include other types of evaluations in addition to the downward (boss-to-subordinate) evaluation. These include self-assessments, peer assessments, and "upward" or management evaluations (employee-to-supervisor); all provide important data for use in the evaluation.

- *Self-assessments*—People are asked to evaluate themselves and outline their strengths and weaknesses. If nothing else, self-assessments provide a good springboard for dialogue between the employee and the supervisor and may help reduce some of the defensiveness that often creeps in when the employee feels "lectured to," given that a downward evaluation is naturally one-sided.
- *Peer assessments*—Members of a group (a department, account, or team) are asked to evaluate their fellow colleagues. In certain service industries, client counterparts might also be polled concerning an employee's performance.
- *Upward evaluations*—Managers who oversee employees should be evaluated on their supervisory ability as part of their overall appraisal. Who best to provide input on supervisory skills than the employees themselves?

A significant part of the appraisal process involves gathering the information and reaching a conclusion regarding the employee's performance. Once the appraisal itself is completed, the appraisal "interview" takes place; this is the critical concluding link of the appraisal process in which supervisor and employee meet to review and discuss the evaluation. Not surprisingly, few people look forward to the confrontational nature of an appraisal interview; both parties often feel nervous, tense, and defensive.

Several factors play a role in how comfortable this meeting will be, including the method of delivery (a supportive supervisor is more effective than one who is threatening), the existing relationship between the supervisor and the employee, and the nonverbal cues that are projected during the interview. In general, though, an interactive mode of communication seems to improve the tenor and outcome of the interview. In other words, giving employees substantial participation in appraisal interviews and conducting two-way discussions increases satisfaction with the interview and sparks motivation to improve subsequent job performance. If feedback has been frequent throughout the interval since the previous performance appraisal, both supervisor and employee should feel more comfortable that there will be no unpleasant surprises.

In addition to a review of past performance and discussion of strengths and weaknesses, the appraisal interview should involve setting job-related plans and goals for the future (an "action plan" for the employee). This action plan should be collaborative—based on the views of both parties and realistic; otherwise, it may become meaningless or even detrimental. (If the employees do not "buy into" the action plan or find the goals set for them impossible to achieve, they may actually lose motivation since their job prospects appear dim.) The action plan is an instrumental tool in evaluating performance during the next evaluation.

BALANCE IN WORK AND PERSONAL LIFE

"A rested employee is essential to a company's business." This statement, from a *Wall Street Journal* article (Joann Lublin, July 6, 2000, p. B4), is one many of us who have worked in corporate America never felt would be uttered or embraced. However, companies globally are beginning to recognize the negative consequences of overworked and overwhelmed workers. Employees need time off the job and vacations to develop a balance between work and private life. Technology seems to simplify our work—and makes it more complex at the same time. Consider that time-saving devices like laptops and handheld devices have doubtlessly helped us streamline our work, but they also enable (even force) us to be connected to our jobs at all times. This constant and ubiquitous connectivity makes balancing work and nonwork time far more difficult than ever; consequently, we devote an entire chapter to this topic in Chapter 3.

ERGONOMIC RECOMMENDATIONS

Both managers and employees alike have many factors to consider when they decide to join or leave an organization, hire an employee, or seek ways to make their work lives more pleasant and rewarding. While each person and employer and job are different, a group of attributes that are culled from the literature can likely satisfy people in their jobs and make them successful and productive.

The most important considerations for employees and managers are given in the following, with separate recommendations for each, along with a further list of recommendations that apply to all of us.

ERGONOMIC RECOMMENDATIONS FOR EMPLOYEES

New Job?

Employees should make a thorough attempt to understand their new job environment when they consider accepting work in any organization. "Environment" includes both the organization itself and the individuals within it.

If you are considering accepting or leaving a job, consider the following:

- Examine the components of a company—like its structure, strategy, and policies—and determine if they agree with your own personality and your personal goals and beliefs.
- Pay particular attention to the culture of the organization; it will greatly influence you at work. Choose a company that fits your own personal style as much as possible.

- Job satisfaction is crucial; it affects our lives outside of work, our mental health, and our physical health as well. Consider both internal and extrinsic factors:
  - Examine your own needs and be as realistic as possible about what they are. Does the job meet your physical needs, like sufficient pay and fringe benefits? What about your emotional needs, like recognition and a sense of achievement?
  - Evaluate your work conditions—the office, the dress code, and so forth. Make sure they are appropriate for and pleasing to you.
- Spend some time with your prospective colleagues and supervisors and determine how you feel about them since you will be with them for many hours at work.
- Remember that we are all different; what is important to you may not be important to someone else. That said, take a look at how you feel about your job compared to how others in similar positions feel about theirs. This may become increasingly important to you since social comparisons are part of human nature.
- Find out if your company offers any programs to increase your skills.

Coping with Stress

Inevitably, you will be exposed to stress. Plan out healthy ways to cope with it if you find it becomes a problem:

- If you are stressed, talk about your feelings to empathetic family members or to trained counselors, take time off, prioritize your job duties to feel less overwhelmed, and enlist the help of others if you are overworked.
- Take advantage of any stress-reduction programs your company offers.

What to do if you are unhappy in your job:

- Talk with your supervisor in a nonconfrontational tone about existing problems. It is important that you propose solutions.
- Explore if remote work or flexible hours are possible for you—and if these types of work arrangements would make you happier at work.
- Check your employment contract to determine your duties and rights at work.
- Consult your employee handbook or company policies and procedures manual if you have questions or concerns about your job duties, vacation, and days-off allotment.
- If your unhappiness stems from workplace harassment by other employees or supervisors, consult your company's workplace harassment policy, if it exists, and act accordingly.
  - If no such policy exists, confront the bully if possible.
  - If you are not comfortable with that, or it does not help, carefully document the behavior.
  - Take the documentation to your supervisor, or the next level up, until you have a resolution.

- Talk with your organization's union steward and social consultant.
- If there is no solution to a problem that bothers you so much that you do not want to work at the organization any more, then look for another position.

ERGONOMIC RECOMMENDATIONS FOR MANAGERS

Poor motivation and lack of job satisfaction can lead to poor performance, absenteeism, and turnover. Not only will satisfied, motivated employees be more productive, but they are also less likely to suffer unduly from stress. Here are some pointers to enhance your employees' work lives:

- Make sure your employee's necessary physical and psychological needs are fulfilled—physical needs are for existence, health, and comfort; psychological needs are those required for a being at ease and feeling rewarded, appreciated, and satisfied at work.
- Remember that people place different weights and priorities on needs, and pay attention to what is important to different employees.
- Work conditions—especially policies and procedures, administration, and supervision—play major roles in employee satisfaction; find out what conditions are important to your staff and try to provide them.
- We work for rewards—but since individuals feel rewarded by different things, offer rewards that are meaningful and valuable to them. This means finding out what people want before structuring rewards and incentive programs.
- Make sure there is a definite link between performance and rewards; one important tool for measuring performance and linking it to rewards is the performance appraisal.
- Human nature compels us to compare ourselves with others, so keep in mind that employees' motivation and satisfaction are also influenced by their comparison of themselves and their jobs with others. This statement has many implications; for example, if salaries of employees vary, the figures should be kept confidential.
- Provide opportunities for growth and learning; this enhances employee self-esteem and reduces job monotony. This includes offering ongoing training programs.
- Be sensitive about workplace harassment; there should be a formal employee policy against workplace bullying, and any reports of bullying should be taken seriously, treated promptly, and handled in complete confidence.
- Increase the amount of thinking and decision-making in an employee's job; this means letting employees self-manage when possible or feasible, letting employees get involved in the overall process of their jobs, and letting them contribute to corporate decisions.
- Make sure performance standards or work output requirements are reasonable; help protect employees from excessive workloads.
- Give employees plenty of breaks and vacations; protect their nonwork lives and downtime.

- Provide opportunities for socializing.
- Make sure work stations are comfortable, well equipped, in proper ergonomic conditions. (Appropriate ergonomic conditions are covered in the following chapters.)

Stress is a fact of work life and can be very detrimental to the health and happiness of your employees. To combat stress,

- Provide a supportive environment for employees, paying particular attention to direct supervisors—they should be particularly supportive.
- Consider offering stress-reduction programs if feasible; these may include time management seminars, wellness programs, and relaxation training.

ERGONOMIC RECOMMENDATIONS FOR EMPLOYERS AND EMPLOYEES

- Understand your colleagues—supervisors and staff—and their context, including their goals, the pressures they face, their strengths and weaknesses, and their work styles.
- Understand yourself, assessing your own strengths and weaknesses and your personal style.
- Incorporate the first two steps and develop and maintain a relationship that fits both your needs and style, meets your expectations, and is based on dependability and honesty.
- Practice active listening when interacting with colleagues, which entails playing close attention to what your conversation partner is saying, responding to the information they give you without leading or giving advice, and identifying your partner's feelings and meaning by absorbing their comments and repeating portions to verify the information.

REFERENCES

Alderfer, C. P. (1972). *Existence, Relatedness, and Growth: Human Needs in Organizational Settings.* New York: Free Press.

Ayu, A. (2014). The enormous cost of unhappy employees Inc. *Magazine*, August 27, 2014. http://www.inc.com/ariana-ayu/the-enormous-cost-of-unhappy-employees.html. Accessed March 10, 2016.

Canadian Centre for Occupational Health and Safety. (2015). Bullying. http://www.ccohs.ca/oshanswers/psychosocial/bullying.html. Accessed March 10, 2016.

Conway, F. T. and Smith, M. J. (1997). Psychosocial aspects of computerized office work. In M. Helander, T. K. Landauer, and P. Prabhu (Eds.). *Handbook of Human–Computer Interaction*. Amsterdam, the Netherlands: Elsevier, pp. 1497–1517.

Greenberg, J. and Baron, R. A. (2000). *Behavior in Organizations* (7th ed.). Boston, MA: Allyn & Bacon.

Hackman, J. R. and Oldham, G. R. (1980). *Work Redesign*. Reading, MA: Addison-Wesley.

Hendrick, H. W. and Kleiner, B. M. (1999). *Macro-Ergonomics: An Introduction to Work System Analysis and Design*. Santa Monica, CA: Human Factors and Ergonomics Society.

Herzberg, F. (1966). *Work and the Nature of Man*. Cleveland, OH: World Publishing.

Hunter, I. (2015). Bullying at work on the rise. *Daily Mail*. http://www.dailymail.co.uk/news/article-3319812/Bullying-work-rise-Helpline-said-received-20-000-calls-employees-year-saying-considered-suicide.html?ITO=applenews. Accessed March 10, 2016.

Huppke, R. The impact of workplace bullying. *The Chicago Tribune*, October 16, 2015.

Kroemer, K. H. E., Kroemer, H. B., and Kroemer-Elbert, K. E. (2001). *Ergonomics: How to Design for Ease and Efficiency* (2nd ed.). Upper Saddle River, NJ: Prentice Hall.

Locke, E. A. and Latham, G. P. (1984). *Goal Setting: A Motivational Technique That Works*. Englewood Cliffs, NJ: Prentice Hall.

Lublin, J. (2011). Memo to Staff: Stop Working, Take a Thinking Day. *The Wall Street Journal*. http://www.southcoasttoday.com. Accessed March 15, 2016.

Maslow, A. H. (1943). A theory of motivation. *Psychological Review* 50, 370–396.

Muchinsky, P. M. and Culberson, S. S. (2016). *Psychology Applied to Work* (11th ed.). Summerfield, NC: Hypergraphic Press.

NIOSH. (2014). Stress…at work. http://www.cdc.gov/niosh/docs/99-101/. Accessed March 10, 2016.

Parade Magazine. (August 21, 2015). Field guide: Millenials versus baby boomers. *Parade Magazine*. http://parade.com/417123/parade/field-guide-millennials-vs-baby-boomers/. Accessed March 10, 2016.

Selye, H. (1978). *The Stress of Life* (Revised ed.). New York: McGraw-Hill.

Willer, P. (June 22, 2015). Millenials will dominate the workforce. *Forbes*. http://www.forbes.com/sites/sap/2015/06/22/millennials-will-dominate-the-workforce-is-your-business-ready/. Accessed March 10, 2016.

3 Work–Life Balance

It's up to us as individuals to take control and responsibility for the types of lives that we want to lead. If you don't design your life, then someone else may just design it for you, and you may not like their idea of balance.

—Nigel Marsh, author

Don't confuse having a career with having a life.

—Hilary Rodham Clinton

There is an ugliness in being paid for work one does not like.

—Anaïs Nin

OVERVIEW

Work–life balance: if you work, it is probably on your mind. Thoughts and concerns about how much you are working and how your work schedule is affecting you and those around you are a side effect of having a job. With changes in technology and the pressure to do more at work, the concept of work–life balance becomes more pressing today than ever. A great deal of research indicates that taking time off work is critical for overall health and well-being, and time off includes short breaks, days off, and vacations. These are not luxuries, but important parts of any healthy lifestyle. Think of it this way: just as your mind needs a respite from its work routine, so do your fingers, your hands, and your wrists. Your back needs a break from the office chair, your eyes need respite from that screen, and your fingers crave some time away from the keyboard. Of course, although most of us believe that a work–life balance is healthy and that time off is necessary to replenish our energy and renew our focus on work, some experts disagree, decrying the importance of work–life balance as a myth.

In this chapter, we explore why work–life balance can be increasingly difficult to find, how this could affect office employees, and how to go about achieving this balance—assuming you want to do so.

Note: You may skip the following section and go directly to the "Ergonomic Design Recommendations" at the end of this chapter — or you can get detailed background information by reading the following text.

THE 40-HOUR WORK WEEK

During the Industrial Revolution, workdays were long—between 10 and 16 hours a day—because factories had to run around the clock to meet production goals. In the 1920s, Henry Ford, founder of Ford Motor Company, established the 5-day, 40-hour work week, and he did so for a surprising reason: he wanted to reduce staff work hours so employees would have time to shop. In fact, he still paid them the same wages he had paid during the longer work week; he wisely believed that potential consumers needed not just the free time but also the available funds to shop.

So the standard 40-hour workweek is actually the carefully considered outcome of profit-maximizing efforts by Henry Ford. Of course, as a fortunate coincidental result, employers discovered that they could actually get more output from people by having them work fewer days and fewer hours. Since then, researchers have continued to study this phenomenon in a variety of modern industries. The result: beyond 40–50 hours per week, the marginal returns from additional work decrease rapidly and quickly become negative. As one example, a study in the *American Journal of Epidemiology* (2008) found that those who worked 55 hours per week performed more poorly on some mental tasks than those who worked 40 hours per week.

HUMAN NATURE AND WORK

Adam Smith, "the father of industrial capitalism," believed that we as people, at our core, are essentially lazy. In his 1776 tome *The Wealth of Nations*, he said, "It is the inherent interest of every man to live as much at his ease as he can" (Constitutional Rights Foundation, 2007). This notion and its reverberations have set the stage for much of how work was framed from that time forward into modern day. Taylorism, which debuted a century after Adam Smith's book was published, was based on the premise that efficiency was the key to profitability and that an individual worker's job satisfaction was not even remotely a factor for consideration. To incite a worker into showing up daily for menial, unrewarding work, compensation was necessary—and the only reason anyone would appear. Consequently, for many decades, employers set up systems of work where compensation is the proverbial carrot and our institutions—factories, offices, shops—are enormous and never-ending hamster wheels of often unfulfilling work.

In 1949, a book by career counselor William J. Reilly, called *How To Avoid Work*, deftly captured the essence of this human desire to find our purpose and do what we love. It is a remarkable work given that it was written at the cusp of what we now picture as Corporate America. Even though women at the time were mostly employed as housewives, the book encouraged people of all ages and genders to pursue their passions in lieu of safe but conventional professions (Popova, 2012). Take a look at these passages:

> Most [people] have the ridiculous notion that anything they do which produces an income is work—and that anything they do outside "working" hours is play. There is no logic to that.
>
> [...] Your life is too short and too valuable to fritter away in work. If you don't get out now, you may end up like the frog that is placed in a pot of fresh water on the stove.

As the temperature is gradually increased, the frog feels restless and uncomfortable, but not uncomfortable enough to jump out. Without being aware that a chance is taking place, he is gradually lulled into unconsciousness. Much the same thing happens when you take a person and put him in a job which he does not like. He gets irritable in his groove. His duties soon become a monotonous routine that slowly dulls his senses. As I walk into offices, through factories and stores, I often find myself looking into the expressionless faces of people going through mechanical motions. They are people whose minds are stunned and slowly dying.

A Gallup poll in 2013 queried 230,000 full-time and part-time workers in 142 countries and found that only 13% of these people felt engaged and fulfilled by their jobs (Crabtree, 2013). This means that almost 9 in 10 people asked are unhappy in their jobs—that statistic is staggering. Keeping in mind Reilly's proposition, we make the point here that this profound dissatisfaction with work is one of our own making—the product of how we have designed our institutions, with income as the primary motivating factor, rather than the actual work itself (whether or not it is fulfilling and enjoyable).

As described in detail in Chapter 2, there are many motivating factors at work and a number of nonmonetary means of reward or compensation. We bring this up here, on the topic of work–life balance, because briefly put, a fulfilling job at 40 hours a week is far preferable—and balanced—than one we hate or find boring.

BALANCE IN WORK AND PERSONAL LIFE

Work–life balance might best be described as work–life *integration*. With today's constant connectivity, margins between being "at work" and being "off the clock" are blurry and ever shifting. Nonetheless, most of us think it is important to disconnect from work; back in the first century, Roman philosopher Seneca the Younger uttered a warning that is as valid today as it ever was: "It is inevitable that life will be not just very short but very miserable for those who acquire by great toil what they must keep by greater toil." We have set a formidable trap for ourselves in our pro-entrepreneurial culture, where it is common not only to accept but actually to praise and admire extreme work schedules.

Americans are among the hardest working people on the planet—or at least they spend the most hours doing it. They work about 1788 hours per year, which may surprise many, given that the famously diligent Japanese work "only" about 1735 hours a year and Europeans average 1400 hours per year (OECD, 2014). A 2014 Gallup poll found that Americans work on average 47 hours per week: in fact, one in five clocks more than 60 hours per week (McCarthy, 2014). With technology ever advancing, we do much of that work in locations other than the traditional office, which allows work to encroach on our "free" time even more. And our constant connectivity and ubiquitous wireless connection gives us more distractions at work as well—shopping online, tweeting about our weekend, and checking personal e-mails.

This is a real problem because such work schedules and the propensity to work around the clock are actually making us sick. The Lancet did research in 2014 that involved a systematic review and meta-analysis of both published and unpublished data for over 600,000 individuals. The analysis showed that employees who work long hours have a higher risk of stroke than those working standard hours. Moreover, the biggest overachievers, working 55 hours a week or more, were 33% more likely to suffer a stroke. The researchers speculated that increased tension could lead to other unhealthy behaviors—drinking heavily, overeating, and selecting nutrient-poor and calorie rich "comfort" food. These behaviors in turn led to poor physical outcomes.

Companies wanting to hire the most promising millennial talent should offer a guarantee of work–life balance and a work environment that appeals to the most entrepreneurial among them (Dill, 2015). This applies particularly to companies in emerging markets, defined as "a country exhibiting a rate of GDP growth higher than the global average, but whose markets are not as mature as those in countries with developed economies."

The desire for work–life balance may be less about time off or the length of the workweek than it is about flexibility and the desire for a creative, dynamic working environment. Global CEO of Universum Petter Nylander (2014) notes that "...this generation, no matter where they [SIC] are in the world, are really taking a different approach to work. Work and life are merging ...and employers need to rise to the challenge and make investing in the work environment a priority." (Dill, 2015). Nylander recommends that companies seeking to hire the best and brightest should offer opportunities for creativity, innovation, and entrepreneurship.

We will make one more argument in favor of work–life balance: Ask yourself how many times you have had an excellent idea, thought of a compelling story line, or came up with an innovative solution to a problem during a routine activity, like while taking a shower or on a morning jog. There is science behind this somewhat surprising fact, as explained by neuroscientist Alice Flaherty (Bourque, 2014). One of the ingredients we need to be creative is dopamine; the more dopamine is released into our bodies, the more creative we are. A warm shower, exercise, and similar activities make us feel relaxed and content, and dopamine flows. Another crucial factor is distraction: something to disengage our brain from fixating on certain topics or issues or moving into a rut. When our minds are distracted, not diligently focused on a given work problem, we tend to look inward, and alpha waves can ripple through the brain, stimulating our subconscious. This is where we start to make connections that we could not make before. This magic combination—relaxed state of mind, distracted, full of dopamine—tends to lead to the best, most creative ideas.

Dustin Moskovitz, who worked on Facebook with Zuckerberg while both were at Harvard, has evocative and compelling words for those intense times and unimaginably long workdays:

> Again and again, I came back to the idea that I wish I had *lived my life* differently. 2006 was one of the best years for Facebook, and one of the worst years for me as a human.

> I wish I had slept more hours, and exercised regularly. I wish I had made better decisions about what to eat or drink—at times I consumed more soda and

energy drinks than water. I wish I had made more time for other experiences that helped me grow incredibly quickly once I gave them a chance.

You might think: but if you had prioritized those things, wouldn't your contributions have been reduced? Would Facebook have been less successful? Actually, I believe I would have been *more* effective: a better leader and a more focused employee. I would have had fewer panic attacks, and acute health problems—like throwing out my back regularly in my early 20s. I would have picked fewer petty fights with my peers in the organization, because I would have been generally more centered and self-reflective. I would have been less frustrated and resentful when things went wrong, and required me to put in even more hours to deal with a local crisis. In short, I would have had more energy and spent it in smarter ways… AND I would have been happier. That's why this is a true regret for me: I don't feel like I chose between two worthy outcomes. No, I made a foolish sacrifice on both sides.

Goudreau (2014)

IS BALANCE OVERRATED?

We would be remiss if we did not mention the dissenting views: some experts disagree about the notion of work–life balance. The argument against work–life balance postulates that work is what we spend the majority of our days doing and what we do for a living becomes a core part of who we are as a person. It is also how we provide for ourselves and others financially. And for those of us lucky enough to do what we actually enjoy for a living, work can be especially enveloping. Add mobile technology into the mix, and work can easily take over our personal lives.

It is hard to argue that for some, work–life balance is just not desirable; some people are simply that driven. Take the examples of successful founders of some of our most innovative international technology companies, like Apple or Facebook. Facebook Chief Executive, Mark Zuckerberg, reportedly spends 50 or 60 hours a week in his office, spending very little energy on anything but work. For example, his practice of wearing the same "uniform" of jeans and a T-shirt every day avoids wasting time spent choosing and shopping for clothing. "If you count all the time I am focused on our mission, that is basically my whole life" (Shandrow, 2015). Under the direction of Bill Gates, Microsoft grew into an enormous success, and Gates was known for massive coding sessions where he would fall asleep on his desk, only to awaken and start typing again right away.

This intense passion that some people possess may be understandable and possibly essential in some instances and under certain circumstances. This is true not just today but also in early times; consider this quote by Larry Keeley, cofounder of Doblin, an innovation consulting agency: "I don't think there was any work–life balance while the Sistine Chapel ceiling was being painted" (McFarland, 2015). The same might be said for Einstein or da Vinci or any number of early inventors or musicians, artists, and writers.

Moreover, not every company "got the memo" on work–life integration. An August 2015 expose on "the USA's most valuable" company, Amazon, demonstrated just how far some companies still go to push their white collar employees to deliver maximum results regardless of the effect on their personal well-being (Kantor and Streitfeld, 2015). According to the *New York Times*, Amazon employees are expected to work long, late hours, endure high turnover and massive annual firings, and receive vocal and blunt feedback in order to keep their jobs. Among Amazon veterans, many said the workloads were often extreme: marathon conference calls on Easter Sunday and Thanksgiving and hours spent working at home most nights or weekends.

Nonetheless, employees willingly accept the working conditions, and we conclude that the work itself must be involving and satisfying enough—combined with compensation and other company perks—for them to remain (and often thrive) in Amazon's employ. Fact is many people very much want to work for Amazon and are delighted at the opportunity of working with smart, like-minded colleagues who care deeply about their work and are thrilled by the opportunity to help create the "next best thing."

RECOMMENDATIONS FOR ACHIEVING WORK–LIFE BALANCE

We again find ourselves in subjective territory; what works well for one individual at one moment in time might not work well for someone else. Even if we can agree that some circumstances call for extreme bouts of work, over time, few people can or should keep up this pace over extended periods of time. Generally speaking, most of us would like to experience a comfortable, happy lifestyle, with a satisfying and fulfilling work life rounded out by an equally satisfying and fulfilling nonwork life.

With exceptions made for necessary work surges, we do believe that work–life integration is important to our physical and emotional well-being. So how do we go about achieving it? Harvard Business School (HBS) drew on 5 years worth of interviews with close to 4000 executives worldwide to compile research in 2014 that uncovered some interesting facets of work–life integration (Groysberg and Abrahams, 2014).

The HBS research highlighted five main themes in achieving that balance: defining success for yourself, managing technology, building support networks at work and at home, traveling or relocating selectively, and collaborating with your partner/family (we meld the latter two together as there is a great deal of crossover between the two). We cover each of these in turn below and add some additional techniques and guidelines for each theme.

DEFINE SUCCESS FOR YOURSELF

This should be done in both the professional and the personal realms. For example, an executive might define one aspect of professional success as managing his clients well and one aspect of personal success, as having an outing with his or her family every Saturday. Importantly, these definitions change over time and with evolving circumstances. Also, Groysberg and Abrahams noted a gender difference in definitions of success, with women mentioning cultural expectations about mothering and emotional guilt of being separated from their children, whereas men still saw their family responsibilities through the lens of breadwinning. It is human nature to define ourselves and derive some self-worth through our jobs, and there is certainly nothing wrong with being dedicated or ambitious, but it is dangerous to cling too closely to work as self-worth.

> *Communicate your needs*: Defining success is the first part of this theme; communicating the resulting needs that go along with these definitions is next. Communicating the needs means explaining them to colleagues, supervisors, family, and friends. Your colleagues, for example, will not know that you have committed to 7:30 Monday morning yoga classes with your spouse unless you tell them, and once you do, they can stop asking for, say, a coffee meeting at that time. Let us assume that you define one of your goals as having dinner with your family three nights a week. If that means leaving the office a bit early on those three nights, you must communicate that to your office managers, colleagues, and staff; they will not know your routine unless you tell them.
>
> *Respect the boundaries you set*: The second step of this overall theme is to then respect the boundaries of the goals you have set. This might be difficult at first, but sticking to the boundaries you have set will develop a routine over time and establish your commitment to the goals.

MANAGE TECHNOLOGY

This is an almost universal concern among the executives interviewed, and we all need to decide on our own when, where, how, and how much to be available for work (Groysberg and Abrahams, 2014). Consider here that always being plugged in can actually undermine performance; as we mentioned earlier in this chapter, there is an actual scientific reason why we are more creative when we are involved in a nonwork routine. Albert Einstein himself noted "I think 99 times and find nothing. I stop thinking, swim in silence, and the truth comes to me." While mobile technology has made life easier in many ways, it has also complicated our ability to truly be away from work.

> *Embrace the off button at home*: The reason work encroaches more and more on our personal time is that we hand over our power to alerts and notifications so we will not miss anything—News, Twitter, e-mails, and posts are often all set to audibly alert us when they arrive. We let ourselves be bombarded with these alerts, and we gradually lose our ability to distinguish between priorities and sheer information. Technology comes with an off button, and we must learn to use that button. If you find that difficult, you are not alone. Consider doing it in phases—leave your cell phone at home when you walk the dog; avoid checking work e-mails on a Saturday.
>
> *Embrace the off button at work*: Let us be honest, most of us do not shut down our personal lives when we are at work. A 2014 Survey.com poll shows the

number of people who admit to wasting time at work every day has reached a whopping 89%, with much of that time being spent on personal use of technology—nonwork-related e-mails, texts, online shopping, and online surfing (Conner, 2015). If we turn our attention to work and dial back the distractions, we will probably find ourselves working more efficiently, and we will accomplish what we need to more quickly, which translates into fewer hours spent at work.

BUILD SUPPORT NETWORKS

Support networks can be formal (i.e., paid caregiver at home) or informal, and they should exist both at home and at work. For the latter, trusted work peers serve as advisors or sounding boards, and a supportive boss or helpful colleagues step in when a crisis at home requires time away from work. Similarly, emotional support at home is extremely valuable, with friends or family helping when an individual needs perspective or assistance. Some ways support networks are helpful:

- *Advising and problem solving*: Even just the process of describing a problem you are grappling with to someone else can help you find a solution.
- *Reducing stress in a crisis*: Have you heard the saying that "trouble shared is trouble halved" (Dorothy Sayers)? People helping you in a crisis, even if just by lending an ear, goes a long way in reducing stress levels.
- *Mentoring*: This is beneficial both to the mentor and to the mentee: the inexperienced person receives the benefit of a seasoned viewpoint, and the mentor learns and grows from the act of teaching.
- *Practical help*: Practical help encompasses a variety of tasks, ranging from babysitting to running errands to expediting a bit of information.

Extending this idea, friends and/or family should be drawn into major work decisions involving relocations or career changes. The HBS research again showed gender disparities here, with women relying more on assistance with child-related tasks and more prone to keeping work and home networks separate.

COLLABORATE WITH PARTNER/FAMILY

Groysberg and Abrahams reported that employees with strong family lives spoke again and again of needing a shared vision of success for everyone at home—not just for themselves but for the whole family. Most of the executives in the HBS sample had partners or spouses, and common goals hold those couples together. Their relationships offer both partners opportunities that they might not otherwise have had, like travel, parenting, continued education, community involvement, or whatever else is of interest. As part of this theme, we include travel or relocation and the value of pacing ourselves:

Traveling or relocating: If travel or relocation are part of your job, or your spouse's or partner's job, these types of decisions are best made as a team—a family unit—with attention paid to what is best for every member of that team. Travel or relocation becomes especially tricky with children at home and can have lasting reverberations on the rest of the family, so travel and relocation should be carefully examined and considered. For the HBS study, many women reported cutting back on business trips after having children, and executives of both sexes said they had refused to relocate while their children were adolescents.

Pacing yourself: You and your loved ones presumably have a long, healthy, productive, and happy life and career ahead, and understanding the value of pace is critical. There are times when you need to throttle up and there are times when you can slow down. With some thoughtfulness and self-awareness and with an eye to how your work life is affecting your loved ones, pacing yourself might just help you enjoy the journey as much as the destination.

SUMMARIZING WORK–LIFE BALANCE

In sum, many companies today recognize that frustrated, overtaxed employees who do not have a satisfying life outside work are less effective in the longer term. To help employees balance their time, some companies restrict weekend and vacation work, while others are implementing sabbaticals and offering personal time. Examples are all around us: over 5% of U.S. companies offer paid sabbaticals (Arndt, 2006, survey by Society for Human Resource Management); eligible McDonald's employees can take an 8-week paid sabbatical for every 10 years of service; some companies have implemented meeting-free days; a town in Sweden is experimenting with a 30-hour workweek for a year beginning early 2015 to gauge productivity levels of a shorter work schedule.

ERGONOMIC RECOMMENDATIONS

In most any career, there are times where we must work long hours and other times where workloads ebb. We all work differently, and ideal working hours are highly subjective. Work–life balance is a self-defined, self-determined state of well-being, one that allows us to effectively manage our many responsibilities, including those at work, at home, and in our community. An effective work–life balance supports our physical, mental, emotional, familial, and community health, with little stress or negative impact.

As an employer, you can help your employees find this work–life balance by establishing the following:

Ensure that human resource policies are family friendly and that they apply to both men and women in all states of career development:

- These policies should evolve to reflect cultural shifts; for example, today, many adults care not only for their children but also for aging parents.
- They should apply to employees at any level of career with the company.

Give employees more control over their time:

- Allowing employees to negotiate flex time and remote work when feasible establishes a better working relationship.

- Employees who feel in control of their time are generally more satisfied at work.
- Giving employees flexibility recognizes the reality that everyone has different needs at different points in their lives.

Dispel the notion of "face time":

- The idea that managers need to "see" their employees to know that they are working and productive is passe and undermines the notion that we trust and value our employees.
- Employees should be held to the same performance targets regardless of where or when they work.
- Performance on these targets should be used to evaluate the employee rather than the time spent in the office.

As an employee, you can take measures to balance work and life by considering the following:

Define success for yourself by setting qualitative and quantitative goals about your work and your personal life:

- Recognize that these goals will change as your circumstances at home and at work evolve.
- Once you've set these goals, communicate them to your work associates and your family and friends, and stick to them as much as possible.
- Examples of such goals include "eating dinner with the family four times a week" or "meeting clients only on Mondays and Tuesdays."

Manage technology rather than letting technology manage you—technology allows us to be connected constantly:

- Make a conscious decision when, where, how, and how much to be available for work during nonwork times.
- Remember that your brain can actually be more creative at times when you are enjoying "down" time.
- Embrace your technological devices' off button at home.
- Consider eschewing the personal business you conduct on your technological devices while at work to shorten your workday—distractions caused by personal business in the office may keep you physically at work longer.

Build support networks at work, at home, and in your community:

- Support networks can be formal—like a paid caregiver or a grocery delivery service.
- Informal support networks include trusted work colleagues that can act as a sounding board for a business decision or a group of close friends who can help in personal crises.

- Support networks can help through
  - Advising and problem solving
  - Reducing stress in a crisis
  - Mentoring
  - Practical help

Collaborate on career goals and decisions with those people who are important to you:

- If travel or relocation is part of your job or your spouse's or partner's job, these types of decisions are best made as a team.
- Pace yourself so that you can enjoy the journey as well as the destination.

REFERENCES

Arndt, M. (2006). Nice work if you can get it. *Bloomberg Business Report*. http://www.bloomberg.com/bw/stories/2006-01-08/nice-work-if-you-can-get-it. Accessed March 15, 2016.

Bourque, P. (2014). Relaxation Rx. [Blog]. http://litcoachlady.com/tag/alice-flaherty/. Accessed March 15, 2016.

Conner, C. (2015). Wasting time at work: The epidemic continues. *Forbes*. http://www.forbes.com/sites/cherylsnappconner/2015/07/31/wasting-time-at-work-the-epidemic-continues/. Accessed March 15, 2016.

Constitutional Rights Foundation. (Spring 2007). Free markets and anti-trust law. 23(1). http://www.crf-usa.org/bill-of-rights-in-action/bria-23-1-a-adam-smith-and-the-wealth-of-nations.html.

Crabtree, S. (2013). Worldwide, 13% of employees engaged at work. *Gallup*. http://www.gallup.com/poll/165269/worldwide-employees-engaged-work.aspx. Accessed March 15, 2016.

Dill, K. (2015). Millennial job candidates in emerging economies seek work-life balance, entrepreneurial environments. *Forbes*. http://www.forbes.com/sites/kathryndill/2015/08/26/millennial-job-candidates-in-emerging-economies-seek-work-life-balance-entrepreneurial-environments/. Accessed March 15, 2016.

Goudreau, J. (2014). Dustin Moskovitz, The second-youngest billionaire in America, discusses what it feels like to be filthy rich. *Business Insider*. http://www.business-insider.com/young-billionaire-dustin-moskovitz-on-being-rich-2014–10. Accessed March 15, 2016.

Groysberg, B. and Abrahams, R. (March 2014). Manage your work, manage your life. *Harvard Business Review* 92(3), 58–66. https://hbr.org/2014/03/manage-your-work-manage-your-life.

Kantor, J. and Streitfeld, M. (2015). Inside Amazon: Wrestling big ideas in a bruising workplace. *New York Times*. http://www.nytimes.com/2015/08/16/technology/inside-amazon-wrestling-big-ideas-in-a-bruising-workplace.html.

McCarthy, N. (2014). A 40 hour work week in the United States actually lasts 47 hours. *Forbes*. http://www.forbes.com/sites/niallmccarthy/2014/09/01/a-40-hour-work-week-in-the-united-states-actually-lasts-47-hours/. Accessed March 15, 2016.

McFarland, M. (2015). Can a company innovate without working its employees to death? *The Washington Post*. https://www.washingtonpost.com/news/innovations/wp/2015/08/28/can-a-company-innovate-without-working-its-employees-to-death/.

Nylander, P. (2014). Can work life balance be a career goal? [Blog]. http://universumglobal.com/blog/2014/09/can-worklife-balance-career-goal/. Accessed March 15, 2016.

Organisation for Economic Cooperation and Development. (2014). www.oecd.org/gender/data/balancingpaidworkunpaidworkandleisure.html.

Popova, M. (2012). How to avoid work: A 1949 guide to doing what you love. *Brainpickings.* https://www.brainpickings.org/2012/12/14/how-to-avoid-work/. Accessed March 15, 2016.

Shandrow, K. L. (2015). Surprise! Mark Zuckerberg isn't a workaholic. Well, not exactly. *Entrepreneur.* http://www.entrepreneur.com/article/245139. Accessed March 15, 2016.

Virtanen, M., Singh-Manous, A., Ferrie, J., Gimeno, D., Marmot, M., Elovainio, M. et al. (2008). Long working hours and cognitive function: The Whitehall II study. *American Journal of Epidemiology* 169(5), 596–605.

4 Office Design

All architecture is shelter, all great architecture is the design of space that contains, cuddles, exalts or stimulates the persons in that space.

—**Philip Johnson, architect**

People ignore design that ignores people.

—**Frank Chimero, designer and author**

OVERVIEW

Many of us who are working in an office will spend about one-third to one-half of our waking hours in that office, so our office's design is important. The interior design of our office affects our quality of life, and it can appreciably influence our productivity as well. Consequently, an organization can help increase its chances of economic success by designing its office space so that it helps its users to function efficiently and effectively. Similarly, if we work at home, or via other remote location, we should be sure our surroundings are conducive to our work and appeal to our aesthetic preferences. This chapter will present a brief history of office design and a discussion on how the office environment affects employee morale and expresses the corporate "personality." (Please note that a more detailed review of the history of office design can be found in Appendix B.) We also cover a range of office design options and discuss the process of designing a new or refurbished office. We conclude with a discussion of the home office and other remote work locations before presenting ergonomic recommendations.

LARGE OR SMALL OFFICES?

Office designs can vary from closed, walled-in individual offices to "open" plans. Open plans include the stereotypical "paperwork factory" setup in a huge room with straight rows and columns of desks and chairs—Figure 4.1 depicts an example of such an old-fashioned layout that reminds us of the concept of Taylorism.

Often, large spaces are subdivided by low partitions into smaller semiprivate "cubicles," which have been the object of much jibing. In spite of their popularity as the target of jokes, if done well, partitioned cubicles can indeed give employees some privacy along with the ability to personalize their space. At the other extreme of office design are the separate, walled-in offices used by just one person or shared by two or three employees. In reality, many offices are hybrid designs; they have some areas that feature open plans, while other departments or floors within the company may have individual, separately contained spaces.

FIGURE 4.1 At first glance, an antiquated mass production factory with the same identical workstations set up in straight columns and rows—but (surprise!) note the computers on the desks.

The appropriateness of a given design depends on two main factors: the kind of work performed in the office and the types of personal demands and requirements that the individual users themselves have. There are advantages and disadvantages to all types of office design.

Proponents of open offices praise cost-effectiveness in terms of construction cost and expenses for air-conditioning—heating, cooling, and ventilating the space. They also like the relatively uninterrupted lines of communication among the employees and the ease of collaborating with coworkers. Those who prefer closed-type designs advocate the privacy and individuality of personal offices, recognize the value of separate spaces for concentration-intensive tasks, and appreciate fewer disruptive factors like extraneous noise.

PURPOSE OF OFFICE DESIGN

Whatever the office size, proper space design should seek to minimize any possible adverse effects of environmental conditions like noise and disruptive climate conditions like excessive heat or cold, noxious or offensive odors, and any other similar distractions (more on office climate can be found in Chapter 10). The space needs to be lit effectively, and employees should all interact well; after all, the office is a place to work and produce what we are paid to do.

In many cases, offices are designed also to convey an image of the organization, to express the "personality" of the company, and to offer pleasing aesthetics. The office's appearance contributes to attracting new employees and also helps retain workers. All of these factors combined should increase employees' job satisfaction and performance. Put another way, a company's biggest assets are its property and its people. Having and designing a space that reflects the company and its appreciation for its staff is a powerful lever in recruitment and retention.

Process of Office Design

The process of office design—how to go about it—involves several important steps. The first step is to analyze the needs of the people who will inhabit the office, including the tasks they will perform, the machinery and equipment they will use, and their preferences and work styles. Based on such analysis, the designer (ideally a team of ergonomists, engineers, or architects) should formulate specific statements of these functional requirements; they guide the actual design. The next step is to identify a range of design options so that practical solutions can be chosen; this reflects the need to compromise in the "real" world since the conceptual ideal is not always possible or practical. Then, these options are evaluated (see Tables 4.1 and 4.2) and final solutions are selected. Finally, the chosen design is implemented and put to use, with plenty of training and support provided for the employees. Note that these steps apply also, in principle, to remodeling or updating an existing office. Employees should be involved in all phases of the process so that their needs are truly met and they do not feel manipulated or ignored.

INDIVIDUALITY AND FLEXIBILITY

A pervasive trend in office space design today is the consideration of individuality and personal direction, now especially feasible due to advances in technology. Current engineering know-how allows us to design almost any environment we wish. Yesterday's technology tied us to one office, even to one spot inside that office, because we were bound by wired phones and computers, cabled machines, and stationary equipment; now, many of us can be flexible in location, work schedules, and habits.

Note: You may skip the following section and go directly to the "Ergonomic Design Recommendations" section at the end of this chapter—or you can get detailed background information by reading the following text.

A SHORT HISTORY OF OFFICE DESIGN

Over the course of the past century, not only has our relationship with work evolved, but the spaces we occupy for work have changed as well. We can trace the history of office design by referring to the then-available technology and the prevailing management style and even by the artistic vision of the architects responsible for office buildings; all of these factors influenced the spaces in which we worked. Here, we will provide just a brief overview of office design's history; for a more thorough discussion, please see Appendix B at the end of this book.

In the early 1900s, office design was all about squeezing as much work as possible from employees with minimal attention to employee comfort or morale. Such drudgery was perhaps prophesied by the word "office" itself, derived from the Latin word for "duty." Well into the 1900s, large office buildings, and the arrangement of the offices inside, still followed the early example of the "Uffizi" ("Offices") building in Florence, Italy. Completed in 1581, it was built in U shape around an open court so that every room had a window to the outside. This way, each room had some natural lighting and ventilation. When temperatures dropped, coal fires and stoves could be lit to warm the rooms, but warmth came at a price: the offices became malodorous and stuffy when windows had to be kept closed.

Design plans that followed shapes such as U, O, E, I, H, and T were used into the 1930s so every room could have an outdoor window, although some mechanical ventilations of the lower floors of tall office buildings had been in use since about 1890. In 1906, Frank Lloyd Wright's Larkin Administration building was completed in Buffalo, New York. The Larkin building was a remarkable departure from the norm at the time because it significantly improved workers' environment; its central space was bathed in natural light and it featured "sealed ventilation," with early air-conditioning and radiant heat.

The Milam building in San Antonio, Texas, was the first to be fully air-conditioned in 1936. In Europe, at about the same time, the "International Style" was born and became popular; architects like Le Corbusier designed large office buildings eschewing traditional ornamental flourishes in favor of glass, steel, and concrete in logical design decisions. These buildings had Gustave Lyon's "regulated air" (l'air ponctuel) together with Le Corbusier's "neutralizing walls" (murs neutralisants) with cooled air circulating between double window panes. In 1932, the PSFS building in Philadelphia was erected in the new style and was fully air-conditioned, while the contemporary Empire State and RCA buildings in New York were not. Designed with a T-shaped tower that allowed a massive amount of natural light and rentable space, the tower incorporated the main characteristics of International Style architecture, with little ornamentation and an emphasis on functionality.

One edifice that really utilized all the new technology was Frank L. Wright's Johnson's Wax Administration building, in Racine, Wisconsin, completed in 1939. It was entirely sealed and fully air-conditioned. It had clerestory windows constructed from bundles of Pyrex glass tubing that provided diffuse light to interiors rooms. The large workspace, called "great room," was beautifully lit with indirect light and very little glare, which resulted in an excellent environment for creative work.

With air-conditioning increasingly used in large office buildings and electric lighting well entrenched, buildings began to sprawl and do away with windows. After the Second World War, air-conditioning in office buildings, even of single rooms—often by noisy window air conditioners—became commonplace in North America. This led to growing concerns about the health effects on employees. Were there consequences due to the lack of natural light (Chapter 8)? What about indoor air quality in offices that are chilly or too warm (Chapter 10)? What happens when we put a large group of individuals together in a room and ask them to interact (Chapter 2)? How will they behave, perform, supervise, accept each other, and do so in ways that makes them happy—and be willing to return for work tomorrow?

Today, offices range from small to large, from windowless to naturally light filled, or from skyscrapers to home based to coffee shop based. You may be working from a midsized office in a metropolis surrounded by elevated trains or from a suburban office complex with stocked ponds and a bowling alley. Office design represents just one of the factors that led you to choose the job you have, albeit probably one of the more important aspects.

PURPOSE OF OFFICE DESIGN

Proper ergonomic design of the office space takes a focused, systematic approach in which use- and user-centered requirements are utterly fundamental to the outlay of the space. In the ideal world, we would approach office space design scientifically, identifying and understanding the needs, capabilities, and limitations of the people who will occupy the space and then applying this knowledge in a systematic way.

User-centered requirements consider the tasks or functions to be performed in the space and the equipment and machinery needs associated with these tasks. User-centered requirements concern the comfort needs and preferences of the workers, aesthetic appeal, and safety considerations. Not taking office design seriously is problematic, because the impact of poor office design is profound. According to an international survey, 84% of office workers reported that their work environment adversely affected them by not allowing them to concentrate, express ideas freely, work well in teams, or decide where in the office to work based on their tasks (Dhanik, 2015).

In the real world, of course, we also must consider budget constraints, size limitations, hierarchies within organizations, technological restraints and realities, and time schedules. In many cases, after all, organizations cannot build anew but instead have finite square footages available to them. Moreover, that available space can be reconfigured only under given restrictions, office work must continue in the meantime, and there are deadlines for office and personnel moves. Acknowledging the real versus the ideal world, the logic and procedure of responsible office design apply to creating new spaces as well as to rearranging existing ones.

OFFICE ENVIRONMENT

Offices inherently expose people to environmental conditions that may affect their health, comfort, their willingness, and their ability to perform. The primary goal of office design is to ensure the best possible conditions for employees, with safety (Chapter 7 in this book) and comfort top of mind. Control of lighting (Chapter 8), sound (Chapter 9), and climate (Chapter 10) are the most important environmental conditions influencing how office inhabitants feel and perform.

Lighting (which is covered extensively in Chapter 8) strongly affects us in our work environments. Human factors engineers can set up lighting in the office so that visual work tasks are easy to do. The lighting engineer's design goal matches our own: we want an office environment that (1) allows us to see what we need to see clearly and vividly but prevents glare and annoying bright spots in our visual field and (2) appeals to us in terms of contrast and colors. Task lighting—a lamp or light source attached to and controllable at the workstation—is also widely adopted: where we once saw overall high levels of overhead lighting, focused spot-lit "islands" are now popular. Task lighting saves energy, is aesthetically appealing, and gives us control over our own light source, which leads to additional individual comfort and satisfaction. A number of companies also recognize the value of natural light and are including glass in skylights and windows to add light and thereby an open, more spacious feel to the workplace.

Sound and its evil twin, noise, pervade our space in a myriad of forms—examples include ringing, beeping, and chiming phones; chirping fax machines, clinking printers, clattering keyboards, and churning printers and copiers; chatty colleagues; a coworker's blaring headphones; and even external commotion from raging traffic, nearby passing trains, or boisterous pedestrians ambling by. We are accustomed to a certain background sound level, often generated by the climate control system, which helps to mask some disruptive noises. Even lack of sound can be disruptive— it can actually be too quiet (see Chapter 9).

Research and personal experiences suggest that noise usually is the most intrusive of the environmental factors affecting people in the office. Accordingly, acoustics play a prominent role when planning office spaces. To help reduce noise, we can include tools like sound absorbent screens, carpets, draperies, and outer shells of filing cabinets in our offices.

Air ventilation, temperature, humidity, and, relatedly, smells and other contaminants also influence our comfort and, in their extremes, our health at work. Consequently, they too have an impact on our productivity, performance, and willingness to spend long hours in the office. In our work environment, the climate—the term is used here in its physical meaning—should be neither too hot nor too cold; neither too damp nor too dry. We also want fresh rather than stale air, but the airflow should not generate a strong draft (please see Chapter 10 for further climate details). The latest office spaces offer unprecedented levels of employee comfort; some even offer individual climate control at each workstation.

OFFICE DESIGN AND EMPLOYEE MORALE

Another function of office space design is to attract good employees, keep them on board, and increase their job satisfaction. Indeed, current initiatives in office design are centered on retaining employees and encouraging them to expand their office hours beyond the usual working hours. With people spending long hours at work, one of the ways to keep them productive is to make them comfortable and happy with their office space. Job satisfaction, after all, affects productivity and performance, and work conditions have been shown to influence the level of an employee's satisfaction on the job. (Chapter 2 discussed links between work conditions and job satisfaction.)

In a nod to the fact that an appealing office design may help attract prospective employees, in some instances, qualified job seekers can look forward to a number of amenities. Many organizations presently involved in interior build-outs are insisting on innovative options to draw personnel. Popular types of amenities focus on convenience and on creating a homelike atmosphere at the office. These features go well beyond the typical employee cafeteria and may include free brand-name coffee and juice bars, fully stocked larders, and fitness centers that feature posh locker rooms with laundry facilities and on-site dry cleaning, childcare, on-site massage, and recreational options. All of these make it easier for people to be—and stay—at the office. Making the office like an extension of the home and keeping it as convenient as possible become especially important when employees' duties require extended hours or overtime.

OFFICE DESIGN AND CORPORATE "PERSONALITY"

The design of an office, and how it is run, express the corporate character or personality of an organization. In a sense, the office serves as a company's "visible motif." A workspace should reflect what the organization cares about, all while facilitating the working styles and preferences of the team.

An office designed to take into account employees' opinions and needs increases creativity, productivity, collaboration, engagement, health, and happiness; conversely, when an office fails to reflect a company's values or personality, that can have a negative impact on employees (Heels, 2015). Put another way, we shape offices and office designs, but they shape us, too. We can take an extreme example to illustrate this: imagine a prison, where there is a complete absence of control for the inmate, a total lack of stimuli, and an environment of misery and darkness. Then, consider how that affects mental health, physical well-being, and how the inmate's personality is shaped over time. The aesthetics of our immediate environment over time can be deep and far reaching.

Imagine that you are running a midsized consumer goods company. If you are hiring an agency to do your company's advertising campaign, when you are seeking a highly creative Cannes contender for your television ads, you might expect to find stylized, irreverent offices stocked with pool tables and cappuccino machines when you visit the agency. On the other hand, if you are hiring a CPA firm to handle corporate taxes and the annual audit, you might prefer retaining a company with a more traditional office look.

RANGE OF OFFICE DESIGNS

As said earlier, office designs can vary from large to small or from "open plans" to single-office spaces for one person. The appropriateness of a given design depends on two main factors: what kind of work will be performed in the office and what types of personal demands and requirements the individual users themselves have.

The "office landscape" is based on an idea that took hold in Germany, and then elsewhere, in the 1960s. Its original objective was to ensure efficiency in the use of personnel and space (Kleeman, 1991). The basic concept of office landscape was to maximize the use of space and facilitate contact among those workers who need to

interact in order to conduct business. The landscaping aspect concerned the arrangement of furniture, office machines, flowers and bushes and small trees to create the
appearance of an irregular, "natural" terrain similar to what one would expect in a
park or large garden. This often resulted in aesthetically appealing office "environments" that, on one hand, took up a relatively extensive amount of space but, on the
other hand, could also accommodate change and allow for expansion as a company
evolved and grew.

In sum, then, the appropriateness of an office layout depends on the type of
work performed, including careful evaluation of the degree of interaction required
and the amount of privacy needed. There are pros and cons to all types of office
design. Advantages of open spaces are that they are generally less expensive than
separate closed offices in terms of building cost, tax write-offs, and utility usage
for lighting and climate control. Moreover, open-plan spaces often facilitate communication and collaboration among workers. Additionally, proponents of open
spaces praise the lack of hierarchy that individual offices may encourage—without
the corner office suites that we may find in more traditional office designs, there
is less of a division between the executives and the staff. Disadvantages of open
designs include disruptive noise—such as speech and equipment sounds—which
may reduce performance and job satisfaction. Noise generated in open offices
distracts a third of the workforce and also is associated with higher amounts of
sick leave taken by employees (Harrison, 2015). The lack of privacy of an open
design may also be an issue as illustrated by the case of Robert and Charlie, in the
following text.

Consider Robert, a lawyer specializing in family law, and his neighbor, Charles,
who runs the local chamber of commerce. Charles works in a small office building just a mile from home, and his company's office occupies the third floor
of the four-floor building. His staff of five occupies the larger of the three rooms
in the office, each in a separate cubicle; his own office is in the smallest room,
and the third room is used as a conference room, as the need arises. His office
door is almost always kept open; his staff collaborates frequently on work that
benefits the community in which they work, and the open floor plan design works
well to facilitate the near-constant interactions. Almost half of all the work that
is completed in this office is done as a group effort.

Robert's office is vastly different from his neighbor's. He is a partner in
a well-known law firm that specializes in family law, and all the partners
and senior legal staff have private offices in their corporate headquarters.
HQ occupies five floors of a sprawling skyscraper in a major Midwestern
metropolis, and each partner has a small team of support staff. In Robert's
opinion, having his private office is an absolute must; he meets frequently with
clients in his space to discuss sensitive and often emotional topics, and he
regularly works on cases that are confidential and private. Very few of his
daily duties are collaborative in nature, and they almost all demand a separate,
personal work environment.

In addition to analyzing the nature of the business, we should also consider individual characteristics and preferences: some people are better able to concentrate on the task at hand even with disruptions, while others can concentrate only when they have complete privacy. Some people are inspired when they share a large table with half a dozen other people, when nobody can "hide" behind a closed door, and when they can simply take their computers to any desk or chair within a large communal space. Others are uncomfortable without the security offered by office walls and feel oppressed by the forced communality of an open area; they prefer the privacy of their own personal space. Some find open spaces liberating; others find them limiting.

Many offices, of course, house a mixture of office designs—wide open spaces here, sections with shoulder-high cubicle dividers there, or closed-off individual offices along the perimeter of the room. In many businesses, tasks that require concentration are interspersed with those that require interaction. The majority of us perform a number of tasks during the typical day, and we generally shift between individual work and work with others that demands team effort.

How Office Space Can Affect Company Culture

A large Minneapolis-based professional services company moved from an older office tower downtown into a newly built-out building a few blocks away. The company, a venerable institution with a rich 60-year history, employed thousands of people; one of the reasons cited by many for its enduring success and impressive growth was its extraordinary corporate culture. Employees often described themselves as part of a "family," with extensive interaction among all levels of the corporate hierarchy and a strong prevailing team spirit. Prior to the move, the company carefully deliberated the decision to relocate. The old offices had sentimental value for many employees; this is where many had begun their careers. However, the new office building was easier to access via public transportation, allowed for increased company growth, and was more reflective of the company's success, with a striking exterior appearance and luxurious interior appointments.

Once the relocation was complete, however, many company members soon regretted it. The move carried with it some unintended and unfortunate consequences. The configuration of the new space was different: in the former building, executives' offices and cubicles were interspersed among staff cubicles, whereas in the new space all executives' offices were located on two separate floors. Initially, the concept behind establishing executive floors centered on easier communications among the directors; additionally, the company wanted to reward executives' performance by providing them with especially posh suites. However, the new configuration sharply reduced the casual interaction between executives and employees that had existed before, when many of the managers followed the strategy descriptively known as MBWA (literally, "management by walking around"). Employees and executives inter-faced daily at the old building, meeting routinely at the coffee machines, in the hallways, and in the cubicles and offices; now, interaction was reduced to business-only discussions at formal meetings. Where before employees wel-comed executives' casual "visits" and drop-ins in their cubicles, inviting the friendly and open banter, they now felt anxious and vaguely frightened during formal business meetings.

After some time in the new space, employees felt disconnected from com-pany leadership, even disenfranchised from the company as a whole. There was now far less of a team atmosphere and more of a "them versus us" philosophy. Interestingly, executives too felt out of the loop with the employees, sensing a formidable new barrier between themselves and the staff. The team spirit that had once prevailed was sharply and unexpectedly curtailed by the office move; an important part of the company culture had been inadvertently and irrevocably destroyed.

THE PROCESS OF OFFICE DESIGN

Theoretically, as pointed out earlier, there are two major issues to be considered in office space design: the work to be done and the employees' needs and prefer-ences. In reality, of course, other factors come into play, like budgets, location, timing, and even the status of people within the hierarchy of an organization. Many of us have worked in companies in which the square footage of individual offices corresponded in direct proportion to their inhabitants' position within the organization rather than to the actual tasks they performed. Yet, conceptu-ally, what should be instrumental in designing office layouts are the tasks that will be done and the technical, physical, and psychological needs of the people who do them.

Office design is task specific. There is no one absolutely "right" design that fits every organization. Yet, a certain step-by-step procedure underlies well-planned designs. The success of the design project relies on first gathering all needed knowl-edge about the work to be done in the new office and about the processes and flow lines by which it will be accomplished. This is similar to the "flow charting" com-monly done by industrial engineers, especially when laying out material handling activities (Kroemer, 1997).

Office design is also company specific: there is absolutely no one "right" solu-tion or one perfect design schematic. Instead, office designs should be planned and created for each organization individually. At the core of an office's design is the analysis of "processes," meaning what really goes on in an office and how work flow is conducted; accordingly, the resulting office design should reflect the actual com-munication needs rather than the boxes and lines of an often quasi-fictional organi-zational chart.

The process of office design consists of several important steps, done sequentially. These steps should be handled by a selected team, ideally composed of its future users, managers, architects, ergonomists, designers, and engineers.

Step 1: The first step involves a thorough examination of what the people who will inhabit the space truly need. What tasks will they perform? What machinery and equipment will they use? Do they have specific preferences and work styles? How much do people need or want to interact? Is there a strict division of tasks? Is there a formal and strictly maintained hierarchy? What are the communication patterns among managers and personnel? We should also consider the company's management philosophy so that the appearance of the office can be fit to the company "personality"—see Chapter 2. Of note here is that staff should be kept informed and involved, as the company moves through these steps.

Step 2: With this knowledge, the team of users and managers, architects, ergonomists, designers, and engineers can formulate specific statements of functional requirements. These statements guide the actual design.

Step 3: Next, the team should identify a range of design options so that practical solutions can be chosen; this step reflects the need to compromise in the "real" world since the conceptual ideal is not always possible. Constraints include time limitations, budgets, and "political" considerations that likely exist.

Step 4: Finally, these candidate solutions are evaluated. For this, a formal process is helpful, with careful definition and selection of criteria by which the candidate designs will be judged. The winning design is the one that garners the overall highest score. The last part of this step involves briefing and including employees to ensure that the entire staff is part of the revamp.

Let us assume that several candidate designs for a new office have been submitted, designated here for short as Office 1, Office 2, Office 3, and Office 4. These candidate designs are to be evaluated by a number of criteria: construction cost (ConstrCost), ability to expand when needed (Expand), running costs (RunCost), appeal to employees and clients (Appeal), expected efficacy and effectiveness of use (EfficEffec), and time that is expected to pass until the office can be occupied (Availability). Of course, there may be different criteria as well, but the following example will indicate how to insert them into the scheme.

The scoring, with 1 as the worst and 10 as the best, is done by a "panel of experts" that includes managers and employees. In our case, the scores given by raters were as shown in Table 4.1.

With all criteria of equal importance, according to the raw scores in Table 4.1, Office Design #2 would be the chosen solution with 38 points, though Design #4 is close with just one point less.

TABLE 4.1

Initial Scores of Four Possible Office Designs

	Office 1	Office 2	Office 3	Office 4
ConstCost	3	1	5	10
Expand	4	10	6	1
RunCost	9	10	1	5
Appeal	7	2	1	10
EfficEffec	8	10	5	1
Availability	1	5	8	10
Total	32	38	26	37

However, generally some of the criteria have higher importance than others. For example, RunCost may be considered most crucial and availability the least. Accordingly, with weights (1–10, with 10 the highest) applied to the criteria, the scores change and end up as shown in Table 4.2.

With weighted criteria factored into the calculations, Office Design #4 becomes the preferred solution, with Design #1 not far behind.

You will notice, of course, that this evaluation procedure can be widely modified; for example, more design solutions may be considered, other criteria applied, and weightings changed. Note also that these steps of office design apply, in principle, not only to the design of a new office but also to updating or remodeling an existing office. It is key that the definition and selection of criteria for judging the candidate designs, and their weighting, eliminate any irrational decision making by the muddled "I like this better." The process outlined is logical and transparent and forces all decision makers to follow the same rules.

Importantly, employees should be involved in all phases of the process so that their needs are truly met and they do not feel manipulated or ignored. If employees know that their inputs are taken seriously, then they will participate and cooperate.

TABLE 4.2

Final (Weighted) Scores of Four Possible Office Designs

	Office 1	Office 2	Office 3	Office 4
ConstCost (×8)	24	8	40	80
Expand (×3)	12	30	18	3
RunCost (×10)	90	100	10	50
Appeal (×7)	49	14	7	70
EfficEffec (×5)	40	50	25	5
Availability (×1)	1	5	8	10
Total	216	207	108	218

It is also critical that offices should be flexible, and we mean that in the literal sense. In fact, many architects and expert designers treat office planning and layout as a continuing dynamic process. Consider the life cycle of a company: most organizations hope to grow. Accordingly, they will need to plan for expanding office personnel and the requisite increasing demands for office space. One approach for keeping office designs flexible is to treat some parts of the office as more "fixed" than others and to keep the variable portions as easy to move or reconfigure as possible. Kleeman (1981) referred to office landscape components as "shell" and "scenery." The building is the shell, which is difficult and costly and time-consuming to change. The scenery consists of the movable and easily changeable elements of interior design like partitions, furniture, and plants. Scenery can be replaced or revamped at relatively short intervals and fairly inexpensively, reflecting the cycles of changing use in an office building.

NEW TRENDS IN OFFICE DESIGN

The trends we see influencing office design today are centered on several primary themes, and we outline them in the following text. The importance of employee well-being and retention, mobile technology, and the idea that work is now a "thing you do" rather than a "place you go" are key drivers of these trends; they are interrelated.

MIXED WORKSPACES

Open office design continues to be popular; currently, more than 75% of U.S. offices have open plans (Bortolot, 2014). The difference between early open design plans and today's open office environment is that office designers are hoping for collaboration between—rather than control of—employees. The goal in design is to foster transparency and offer variety in where employees can "alight" to work; a sort of office-within-an-office idea. Different generations and differing personalities work more productively in various ways, and with the rise of virtual employees and remote work schedules, many employees are out of the office some or much of the time and back in the office at other points of the day. Certain types of work require collaborative relationships among employees, while others need more private work areas, and these functions probably coexist within the same office.

Accordingly, what we are seeing now is a mixture of workspaces, all under one proverbial (or literal) roof: open seating areas, team workspaces, shared workplaces, huddle rooms, activity space, recreational space like game rooms, refreshment areas like cafes, and private offices.

"Touchdown" spaces are increasingly popular; these are workspaces that employees use who spend much of their workweek elsewhere. While the cubicle is still very much present, it is in decline, and it is morphing into different kinds of spaces altogether.

FLEXIBILITY AND MOTION

Many designers are espousing "activity-based planning," which means address-ing office design based on what people do. A decade ago, a given employee may have come into the office at 8, filled up the coffee mug, and moved to his or her cubicle-enclosed work station to check e-mails and phone messages. Today, the person may have checked e-mails during the train ride in, exchanged e-mails via handheld device with another colleague, and by the time he or she arrives at the office, he or she is ready to meet informally with the colleague in a shared workspace to discuss a client project that is due later this week.

Offices are increasingly designed to be "elastic" so companies can expand, contract, and reconfigure the space as desired. Using lightweight materials, room dividers, and similar tools for office reconfigurations lets the office morph easily.

Office furniture is also becoming similarly adaptive, with workstations making it easy to change from sitting to standing as wanted (yes, "sitting" for prolonged periods in 2015 is being called "the new smoking") and back again. Workplace furniture will have even more emphasis on ergonomics, and room dividers, furniture, and other office pieces can be made adaptable and built with noise-absorbing materials.

RISING COSTS

The cost of commercial real estate continues to rise, and companies are reducing the amount of office space per person. By 2017, many North American companies are expected to allocate, on average, only 100 ft^2 of office space per person, down from 175 ft^2 in 2012 and 225 ft^2 in 2010 (PR Newswire, 2012). Cost containment is one of the main reasons for this decline, coupled with the growing use of collaborative space. U.S. businesses expanded into more office space in 2014 than they have in 8 years while rents increased in 70% of the country (PR Newswire, 2015).

Even companies on sound financial footing include increased efficiency of office space utilization in their business plans. And they do this through greater efficiency, smaller work areas and fewer individual offices, and technological improvements—electronic storage instead of filing cabinets, for example.

The challenge here becomes managing real estate costs without compromising comfort and well-being of employees. Companies are implementing office designs that encourage energy efficiency solutions—natural light, multipurpose furniture, and multipurpose spaces.

SUSTAINABLE DESIGN AND ENVIRONMENTAL CONSIDERATIONS

More recommendations and regulations are in place than ever before to ensure environmental best practices, including cutting down on paper (and other) waste, using recyclable materials, purchasing sustainably sourced equipment, using reclaimed furniture, and choosing environmentally safe cleaning products. As data storage becomes increasingly digital and cloud-based, fewer on-site storage areas are required.

Similarly, but separately, bringing nature inside is also becoming popular in office interior design—a friendly trend seemingly flowing from the long-ago concept of the "office landscape."

HOME OFFICE

Following your individual interests and preferences in the layout of your workspace is probably most easily accomplished in your home office—the personal workstation away from the company building. At home, we are free to set up a shop wherever we would like, whether it be in a sumptuously appointed office in a wing of your home or at the kitchen table surrounded by the remnants of breakfast.

Many of the tasks comprising "office work" are now done away from the traditional office within the company building. We now often work on a mobile wireless computer, laptop or handheld, or with a phone, wherever we happen to be, frequently at times other than the old 9-to-5 office hours and increasingly from home. What may at first just have been the occasional "touchdown" spot for some quick office work can easily develop into a habitual workspace.

Telecommuting and working at home are more and more common, yet the term "home office" still has the connotation of a corner in the den, a spot in the kitchen or, at best, a spare room. That inferred meaning also implies the use of furniture that is not specifically designed for office work, like a folding chair at a card table or an easy chair in that corner of the den. However, as the occasional work changes into schedules of workdays with long hours of effort, the room or space housing the home office should follow the ergonomic design recommendations we mention throughout this book.

Depending on one's specific work at home, the workspace must accommodate various tasks: that may be drafting plans, generating samples, creating web content, storing materials, filing documents, and shelving literature. You might run a complete business from your home office, with all the related tasks of organization and administration, bookkeeping, texting, posting, tweeting, e-mailing, and telephoning. Alternatively, you may just need room for a notebook computer or touchpad to check on a project. In any case, the home office deserves to be carefully selected, cooled or heated, properly insulated and protected from noise and interference, and well lit. Of course, your individual needs and preferences are your primary considerations before you begin designing your space.

Chapters 5 and 6 of this book provide information about computers and furniture for the home office, and Chapters 8 through 10 discuss aspects of lighting, sound, and climate; these should be weighed and considered as well.

ERGONOMIC DESIGN RECOMMENDATIONS

When designing office space layout, take a systematic approach:
Identify and understand the needs, capabilities, and limitations of the people who will occupy the space:

- Carefully consider the work itself. It may include confidential or highly focused tasks that require privacy, or it may require face-to-face contact with other people.
- Take into account "real-world" constraints like size limitations of the space, budget constraints, hierarchies that must be observed, and time schedules.

Specific environmental conditions that you should address in designing office space include the following:

- *Noise*: Consider the acoustical requirements in offices; we should be able to perform normal office work with minimal disruption or distraction from standard office sounds like another worker's voice. Acoustically efficient screens, carpets, draperies, and outer shells of filing cabinets can all help absorb noise.
- *Lighting*: The office environment should allow us to see clearly and without glare; task lighting is often the most suitable; we should also consider natural light when feasible.
- Climate control (especially of temperature and humidity) influences our well-being and performance.

Here are the other considerations:

- Since appealing office designs may help attract employees, plan in advance for amenities the company may wish to offer.
- Consider also what the office's design expresses about the corporate character or personality of an organization.

To create an original design or redesign an existing office, we recommend following these steps to establish the most effective and efficient design:

1. Analyze the needs of the workers who will inhabit the office, the tasks they will perform, and the machinery and equipment they will use.
2. Examine the workers' preferences and work styles; this includes looking at the amount of interaction and the level of bureaucracy in the business as well as examining the company's management philosophy.
3. Formulate specific statements of the functional requirements you determined in the first steps; these statements will guide the actual design.
4. Identify a range of design options so that practical solutions can be chosen.
5. Evaluate the possible design options by a set of defined criteria.
6. Select the final design.
7. Implement the design, providing plenty of training and support for employees as you go.

Keep in mind these important guidelines as you plan the office layout:

- Involve employees in all steps of the process so that their needs are truly met and they do not feel manipulated or ignored; you can do this by soliciting their input, asking for their opinions, and giving them any training they need in the new space or on equipment. Ideally, users of the new space would actually be codesigners.
- We highly recommend making the office space as flexible as is feasible; the organization will likely change in the future and require plenty of

variability in space and its layout; one way to accomplish this is through modular panel systems and movable dividers.

- Remember that flexibility extends to furniture; select items that serve several functions at once and can multitask; furniture too should be adjustable to fit any number of potential users.
- Remember that flexibility also extends to power and communications sources, which should also be easily movable, unless the office and its employees are "wireless."

REFERENCES

Arnold, D. (1999). The evolution of modern office buildings and air conditioning. *ASHRAE Journal* 41(6), 40–54.

Bortolot, L. (2014). Designing a better office space. *Entrepreneur.* http://www.entrepreneur.com/article/235375. Accessed March 10, 2016.

Dhanik, T. (2015). Five office design strategies that will make employees happier. [Blog]. http://www.fastcompany.com/3045981/work-smart/5-office-design-strategies-that-will-make-employees-happier. Accessed March 10, 2016.

Harrison, K. (2015). Best practices for setting up your office to maximize employee productivity. *Forbes Magazine.* http://www.forbes.com/sites/kateharrison/2015/06/25/a-new-infographic-offers-best-practices-for-setting-up-your-office-to-maximize-employee-productivity/. Accessed March 10, 2016.

Hedge, A. (2000). Where are we in understanding the effects of where we are? *Ergonomics* 43, 1019–1029.

Heels, L. (2015). Does your workspace reflect your culture? [Blog]. http://chicagocreative-space.com/does-your-workspace-reflect-your-culture/. Accessed March 10, 2016.

Kleeman, W. B. (1991). *The Challenge of Interior Design.* New York: Van Nostrand Reinhold.

Kroemer, K. H. E. (1997). *Ergonomic Design of Material Handling Systems.* Boca Raton, FL: CRC Press/Lewis Publishers.

Lovine, J. V. (2000). Playful ideas infuse modern interior design. *Chicago Tribune.* Commercial Real Estate Supplement.

PR Newswire. (2012). Office space per worker will drop to 100 square feet or below. http://www.prnewswire.com/news-releases/office-space-per-worker-will-drop-to-100-square-feet-or-below-for-many-companies-within-five-years-according-to-new-research-from-corenet-global-140702483.html. Accessed March 10, 2016.

PR Newswire. (2015). Demand for office space highest since 2006. http://www.prnews-wire.com/news-releases/us-office-sector-demand-for-office-space-highest-since-2006-300017267.html. Accessed March 10, 2016.

Raymond, S. and Cunliffe, R. (2000). *Tomorrow's Office.* London, U.K.: Taylor & Francis.

Saval, N. (2014). Cubed, A Secret History of the Workplace. New York, NY: Doubleday.

5 Chairs and Other Furniture

Some days, the best thing about my job is that the chair spins.

—**Unknown**

A good stance and posture reflect a proper state of mind.

—**Morihei Ueshiba, martial artist and founder of the
Japanese martial art Aikido**

OVERVIEW

Ever since individuals began working in offices, like in the Uffizi mentioned in Chapter 4, one "constant" in the office was—and is—*change*. Conditions for workers changed from those stuffy inner chambers, unlit and dank, where workers stooped over their tasks and emerged from a day of toil with sallow skin and gloomy expressions, through the "Mad Men"-esque era of corner offices and secretarial bays, and then to the cubicle landscape that was intended to provide privacy and personal space, while also allowing for relatively easy and open communications.

Office design today is evolving again, presently steering away from cubicles and back to an open office plan, but now with mixed-use spaces and amenities that reflect our nonwork lives. In fact, about 70% of offices in the United States have low or no partitions (Kaufman, 2014).

How a given company designs its office depends on several factors, including the type of work the organization does—that is, does it require privacy, or does it encourage collaboration? But other and probably most important considerations appear to be technology—the size and shape of it—and cost concerns. Consider the computer: when it first started to appear in offices, it was massive and needed its own room. Then it shrank and fit on an individual's desk, and the employee became tied to the computer. Now, the mobile device seems tied to the person.

We hope that "modern" offices will never again resemble the stuffy inner chambers in the earliest of offices, and we further hope that the person working there has, at minimum, a comfortable chair and the ability to move around and take breaks as needed. Even with all of the design knowledge we have today, and have had for many years, another "constant companion" in the workplace over the past centuries was—and still is—employee *discomfort*. Office workers have been struggling with some degree of pain and discomfort for as long as offices have existed.

In this chapter, we review theories of healthy postures and techniques for assessing comfort (and discomfort), we present guidelines for workstation design to maximize comfort, and we delve in depth into that bastion of office work—the office seat.

THE LAST 300 YEARS

Life for an office worker in the late 1800s and the early 1900s, while certainly less physically exhausting than that of a factory worker, was rife with drudgery. Herman Melville described the life of the mid-nineteenth-century men as "pent up in lath and plaster—tied to counters, nailed to benches, clinched to desks."

Back then, as today, technology dictated office design (and furniture design)—consider how much an office could shift shapes once gaslight replaced candlelight, for example—and by the end of the twentieth century, the quill was replaced by the typewriter, just as in subsequent years the abacus was replaced by calculators and computers. Typewriters and their companions, vertical file cabinets, began to be widely used and became commercially successful around 1900. Interestingly, at that time it was common to stand while working; then the concept changed and sitting in the office became customary. The rolltop desk of the late 1800s and early 1900s had functional compartments and pigeon holes to handle the burgeoning volume of paperwork; typewriter desks proliferated to accommodate typists; card-filing desks became popular with the debut and dissemination of the Dewey Decimal System in 1876.

Yet even as office work evolved and office furniture changed with this evolution, office workers throughout the past decades, even centuries, experienced pain—low back pain and musculoskeletal overexertion, often together with eyestrain, were common complaints of office personnel since office work debuted.

According to the American Chiropractic Association, more than half of all working Americans claim to suffer from back pain each year, and this is largely attributable to poor workstation posture, caused in part by improperly fitted furniture (ANSI, 2008). The work force is also dealing with eyestrain caused by insufficient lighting and wrist pain from improper keyboard and mouse usage. Clearly ergonomically appropriate systems are a critical component of a safe and healthy work environment.

The fact that generations of office workers have endured physical discomfort at best and permanent injury at worst, as a result of uncomfortable and unhealthy work conditions, is a serious, disappointing reality that makes ergonomists cringe (see Figure 5.1). Worst of all is that these injuries are completely, utterly avoidable.

Around 1700, Ramazzini, an Italian physician who could be called the "father" of human factors engineering because of his early work on occupational diseases, noticed laborers' poor postures. In 1713, he wrote that workers—he was thinking of tailors—who sat still, stooped, looking down at their work, became round shouldered and suffered from numbness in their legs, lameness, and sciatica. Saying "all sedentary workers suffer from lumbago," Ramazzini advised that workers should not stand or sit still, but move the body and "… take physical exercise, at any rate on holidays" (Wright's 1993, translation, pp. 180–185).

FIGURE 5.1   Klockenberg's 1926 illustration of a typical typewriting working posture.

In the 1880s, body posture became a serious concern to physiologists and ortho-pedists. In their opinion, the upright (straight, erect) standing posture was balanced and healthy, while curved and bent backs were unhealthy and therefore had to be avoided, especially in children. Consequently, "straight back and neck, with the head erect" became the recommended posture for sitting and, in what was deemed logi-cal at that time, seats were designed to bring about this upright body position; office furniture was designed for "sitting up straight." (Of course, there were exceptions for high-level executives who might have a luxurious armchair with plush upholstery.) This idea of "healthy" sitting with back straight, thighs horizontal, and lower legs vertical lasted—surprisingly—for about 100 years.

The truth is our bodies are meant to move around. We are not designed to main-tain any single body position, unchanged, over extended periods of time. This includes sitting still for hours on end. We want to get up and move around, but our stationary computer is tying us, quite literally, down to our workstation: through our hands, which must operate the keys, and through our eyes, which must view screen, text, and keys. Given that our eyes and hands are tethered, our whole body is stuck, riveted to the workstation. We try to ease the effort of maintaining our rigid posture by sitting on our seat as comfortably as we can, shifting and slouching as needed, but our body is beginning to feel miserable.

Office work typically involves finely controlled activities, like keyboarding or writing, and sitting at a workstation to do these activities makes sense. Sitting keeps the upper body stable, supports the body at its core, and requires less muscular effort than standing, especially when maintained over long periods of time. However, the seat must be supportive, feel comfortable in combination with the other office furniture and equipment, and be appropriate for the work tasks.

New work duties, the widespread use of easily portable computers, and changing attitudes mean it is time to rethink traditional design recommendations for office furniture. Ideally, furniture would accommodate a wide range of body sizes, varying body postures and diverse activities; it would assist in task performance, facilitate vision, and allow interaction with coworkers; it would be aesthetically appealing and help make people feel well in their work environment.

Note: You may skip the following part and go directly to the "Ergonomic Design Recommendations" section at the end of this chapter—or you can get detailed background information by reading the following text.

"GOOD" BODY MOVEMENTS AND "GOOD" POSTURES

Every day we experience the discomfort of maintaining our body posture over extended periods of time. Even when we are resting in bed or reclining in a comfortable easy chair, we will need to reposition ourselves after a while and move around. This need to "shake a leg" becomes urgent after we have forced our bodies into a still position for a period of time—during a long opera performance, during a long seated meeting, or standing still "at attention" in a military formation. Our body is quite literally built to move around, not to stand nor to sit still. Consequently, we should design our workstations for movement, but instead, we have been designing them for rigidly seated individuals for about a century now.

It was only a 2-hour flight, but sitting in the back half of the airplane, in the middle seat, with barely any legroom and a reclining passenger in the seat in front of his, the man in seat 41B was getting antsy. He could not wait to get up and stretch his legs. The plane landed—smooth, easy—and began taxiing to the gate, but Amir knew he had to "remain seated until the seatbelt light is turned off." The second it did, go off, he was up, flexing his legs, unfolding himself from his seat, forced into a bending position by the overhead bin, but nonetheless, *up.* His neighbors in the seats next to and in front of him groaned inwardly; rolled their eyes ever so slightly. It would be at least another 10 minutes until they would be deboarding; after all, the flight was full, and all the passengers in the front rows would take their time retrieving their belongings and making their way off the plane. Why stand up? Amir did not care, though. His legs needed stretching; he would happily stand, even bent over, just to make that happen.

THEORIES OF "HEALTHY" POSTURES

In the late nineteenth century, body posture was of great concern to physicians, physiologists, and orthopedists. Orthopedist Franz Staffel reported that German farmers and laborers often had abnormal back curvatures: their spinal columns were either too flat or overly bent: lordotic (backward), kyphotic (forward), and scoliotic (sideways). Staffel and his contemporary orthopedic experts declared these back curvatures unhealthy, hypothesized that they were due to malnutrition and bad postural habits, and concluded that they had to be avoided, especially in children.

Staffel observed individuals standing easy and unconstrained, head high and looking straightforward, and, in a nod to this stance, he classified standing postures according to their spinal curvatures in 1889. He characterized a "normal" posture by a straight spine in the front (or rear) view. Seen from the side, a plumb line from the top of the cranium passes through the cervical vertebrae, the shoulder joint, and the lumbar vertebrae; below, the plumb line falls just behind the center of the hip joint and then down the leg where it stays slightly in front of the centers of the knee and ankle joints. This posture, usually called "upright" or "straight," appeared balanced and healthy to him—and then to his disciples—for a century.

Upright Posture

You may have heard the term "neutral posture" used to describe a presumably desirable position; MedicineNet (2012) defines it as "the posture when the joints are not bent and the spine is aligned and not twisted." We think that there can be some confusion in that term—does "neutral" suggest that all tissue tensions around a joint be balanced, so that the position is stable? Does it imply a minimal sum of tissue tensions (torques) around a body joint? Minimal joint discomfort? A relaxed posture? A posture instinctively assumed for a task, to generate maximum body strength, or to avoid fatigue? Consequently, we use the term "straight" to describe the posture described earlier.

Sitting Upright

In 1884, Staffel published his theories about "hygienic" sitting postures. He recommended holding trunk, neck, and head erect, with only slight bends in the spinal column in the side view. His recommendation for an ideal back posture when sitting was similar to what he advocated for standing: straight up.

Staffel and his contemporaries were particularly concerned about the postural health of children; they felt that school seats and desks should be designed—and the children urged to maintain—that "normal" erect posture of back, neck, and head. Starting in the late 1880s, many often elaborate and adjustable "hygienic and healthy" designs for school furniture were proposed: seats, desks, and seat–desk combinations were all laid out to promote the upright posture (Bonne, 1969; Bradford and Byrne, 1978; Bradford and Lovett, 1899; Merrill, 1995; Tenner a and b, 1997; Zacharkow, 1988). Figure 5.2 shows Schindler's 1890 design of school furniture.

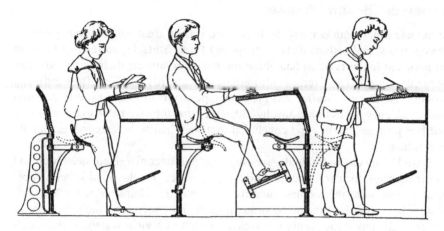

FIGURE 5.2 Schindler's 1890 design of school furniture. About everything that can be thought of for body support in semisitting, sitting, or standing at work.

The sitting posture recommended for pupils was also advised for adults: head, neck, and trunk upright, upper arms and lower legs vertical but forearms and thighs horizontal. For much of the twentieth century, office furniture that forced individuals into this posture was endorsed; even the 1988 ANSI/HFS Standard 100 used it to proscribe office furniture—see Figure 5.3. Interestingly, even as that sitting posture with 90° angles in the joints of ankles, knees, and hips was mostly expected from lowly office personnel, executives and managers habitually enjoyed an ample armchair, upholstered, with a high backrest made to recline for comfortable relaxation. Imagine the difference of sitting hours on end in the sustained stilted upright position versus lounging in an easy chair that literally rocks!

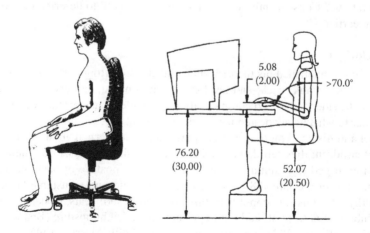

FIGURE 5.3 Unrealistic images of sitting postures circa 1988.

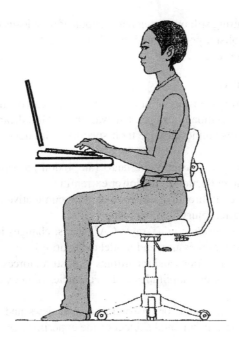

FIGURE 5.4   Hands on keys and eyes on screen.

Designing for this postural ideal completely disregards that it is healthy to change among various postures while sitting, rather than to stay in any one position. Moreover, the typical design of the conventional computer workstation also implies one long-maintained posture: the prevailing idea of "hands on the keyboard, eyes fixated on computer screen" is still widespread. The ensuing positioning of hands and arms, eyes, and head—all fixed in place as depicted in Figure 5.4—determines the position of the upper body and allows very little variation in posture.

Imagine the physical freedom we would have if a keyboard were not needed or if at least it were not stationary and if the display were large and moveable: such design features are technically feasible. There is more discussion of these aspects in Chapter 6 (and in many other publications) but even now it is intuitively clear that no "one and only healthy posture" exits for which the workstation should be designed.

ASSESSING "SUITABLE POSTURES"

As described earlier, for about a century, it was generally accepted that standing or sitting with a straight back was physiologically desirable and socially proper. We could go back to even earlier times and find evidence of accepting this stilted posture: Egyptian tomb reliefs illustrate that the dignified Egyptian man was expected to sit with a straight back, horizontal thighs, and vertical lower legs. There is, of course, nothing wrong with sitting or standing upright at one's own will, but it should be our choice and not be forced on us because of workstation design. Our bodies are made to move, and sitting still over extended periods of time leads to tissue compression, reduced metabolism, poorer blood circulation, and accumulation of extracellular fluid in the lower legs.

Quantitative Measures

The effects of managing a posture or of postural changes can be measured and evaluated by observing variations in dependent variables (Grandjean, 1997; Kroemer et al., 2001). Several disciplines provide their specific techniques:

- *Physiology*: Oxygen consumption, heart rate, blood pressure, electromyograms, and fluid collection in the lower extremities
- *Medicine*: Acute or chronic disorders including cumulative trauma injuries (these are all "after injury" assessments)
- *Anatomy and biomechanics*: X-rays; CAT scans; changes in stature; disk and intra-abdominal pressure; and model calculations
- *Engineering*: Observations and recordings of posture; forces or pressures at seat, backrest, or floor; amplitudes and frequencies of body displacements; and "productivity"
- *Psychophysics*: Structured or unstructured interviews and subjective ratings by either the experimental subject or the experimenter

While most of these techniques are well established, some are not easy to use or are not sensitive to postural variations. Most require a laboratory setting, and few are suited for field studies. For practically all outcomes, however, the threshold values that separate "good" from "bad" conditions are unknown or variable, so interpreting results of those measurements is dicey.

Qualitative Measures

Subjective judgments presumably encompass all the phenomena addressed in the various measurements listed, and they can be reliably scaled and interpreted (Booth-Jones et al., 1998). Based on initial work by Shackel et al. (1969) and Corlett and Bishop (1976), a large number of questionnaires have been developed for assessing people's feelings about uncomfortable or painful working conditions. An often used, well-standardized inquiry tool is the "Nordic Questionnaire" by Kuorinka et al. (1987), modified by Dickinson et al. (1992). It consists of two parts, one asking for general information and the other focusing specifically on low back, neck, and shoulder areas. It uses a sketch of the human body, divided into nine regions—see Figure 5.5. The interviewee indicates any symptoms that may exist in these areas. If needed, more detailed body sketches can be used (van der Grinten and Smitt, 1992). These questionnaires provide information about the nature of problems and complaints—pains and discomfort, their duration, and their prevalence.

Comfort versus Discomfort

Comfort (as related to sitting) has long been defined, simply, conveniently, and misleadingly, as the absence of discomfort. However, Helander and Zhang showed in 1997 that, in reality, these two aspects are not opposite extremes on a single

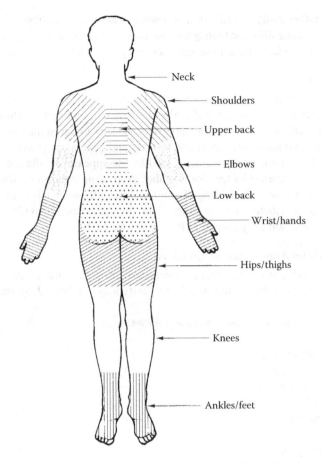

FIGURE 5.5 Body sketch used in the Nordic Questionnaire. The subject looks at the drawing while identifying painful spots.

judgment scale. Instead, there are two scales, one for the desirable feelings of "comfort" and the other for such unpleasant experiences as being ill at ease, fatiguing, straining, aching, and hurting: all terms that indicate some degrees of "discomfort." These two scales partly overlap but are not parallel.

To avoid confusion, instead of using the term "discomfort" in this text, we will employ the term "annoyance" as the descriptive label for the scale containing the unwelcome statements. The other scale, containing the desirable statements, will be labeled with the term "comfort," as has been the convention.

Annoyance Scale

Feelings of annoyance are expressed by such words as *stiff, strained, cramped, tingling,* and *numbness or not supported, fatiguing, restless, soreness, hurting,* and *pain.* Some of these attributes can be explained in terms of circulatory, metabolic, or mechanical events in the body; others go beyond such physiological and biomechanical phenomena.

Users can rather easily describe design features that result in feelings of annoyance such as seats in wrong sizes, too high or too low, with hard surfaces or sharp edges; again, we note here that avoiding these mistakes does not, per se, make a seat comfortable.

Comfort Scale

Feelings of comfort when sitting are associated with such descriptive words as *warm*, *soft, plush, spacious, supported, safe, pleased, relaxed*, and *restful*. However, and importantly, exactly what feels comfortable depends very much on the individual and his or her habits, on the environment and task at hand, and on the passage of time.

Additionally, aesthetics play a role: if we like the appearance, the color, and the ambience, we are inclined to feel comfortable. Appealing upholstery, for example, can strongly contribute to the feeling of comfort especially when it is neither too soft nor too stiff but distributes body pressure along the contact area and if it "breathes" by letting heat and humidity escape as it supports the body.

Ranking Seats by Annoyance or Comfort

Helander and Zhang used six specific statements about seat annoyance or comfort (each with nine steps from "not at all" to "extremely") followed by one general statement.

The statements for *annoyance* were as follows:

1. I have sore muscles.
2. I have heavy legs.
3. I feel uneven pressure.
4. I feel stiff.
5. I feel restless.
6. I feel tired.
   I feel annoyed.

The following statements characterized *comfort*:

1. I feel relaxed.
2. I feel refreshed.
3. The chair feels soft.
4. The chair is spacious.
5. The chair looks nice.
6. I like the chair.
   I feel comfortable.

Helander and Zhang noticed that it is more difficult to rank seats by characteristics of annoyance (as opposed to comfort) because the body is surprisingly adaptive, except when the subject has a sore or injured back. In contrast, comfort descriptors proved to be useful and discriminating for ranking seats in terms of preference. The researchers' subjects said it was easy to make an overall statement about the seats in terms of overall comfort or annoyance (the final answer in the lists) after having responded to the more detailed statements 1 through 6.

We also note that preference rankings of seats could be established early during the sitting trials; they did not change much over the duration. However, it is not clear whether a few minutes of sitting on seats are sufficient to assess them or whether it takes longer trial periods.

"Free-Flowing Motion"

In his 1987 book *Sitting On The Job* (p. 12), Donkin asked: "If slumping in my chair is so bad for me, why does it feel better at times?" The answer lies in the fact that slumping and other body movements are the instinctive attempts to take strain and tension away from muscles that are working to maintain prolonged postures.

In 1984, Grandjean and his colleagues found that individuals in offices often leaned back in their seats even though the seats were not designed for this. Similarly, Bendix et al. (1996) reported that individuals who were reading often assumed a kyphotic (forward) lumbar curve even when sitting on a seat that had a lumbar pad that should have produced a lordosis (backward curve). Evidently, our bodies will lead us to sit in any manner that we find comfortable, regardless of how experts think we should sit.

Let us take a closer look at this hypothesis. In 1962, Lehmann demonstrated the contours of five people "resting" under water where the water fully supports the body. Sixteen years later, NASA astronauts were observed as they relaxed in space. The similarity between the postures under water and in space is remarkable, as shown in Figure 5.6. It appears that in both cases, the sum of all tissue torques around body joints has been nullified. It was not by accident that the shape of the so-called easy chairs is quite similar to the contours shown in both figures.

Our favorite term for current office seat design is "dynamic," meaning that design should recognize and reinforce free-flowing motions, rather than "static" maintained posture. People move around and should be able to take on any posture they wish, as sketched in Figure 5.7.

The "free-flowing motion" or "floating support" design idea has these basic tenets:

- Allow the user to move freely in—and with—the seat and opt for a variety of sitting postures at will, each of which is supported by the seat, and to get up easily and move around when desired.

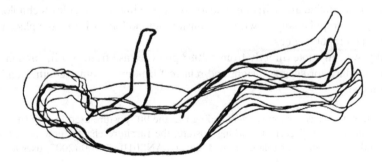

FIGURE 5.6 Relaxed postures under water and in low gravity.

FIGURE 5.7   "Free-flowing motions." People differ in size and preferences, and everybody changes, moves, or gets up—does anybody ever sit still for long periods of time?

- Make it easy for the user to adjust the seat and other furniture, especially keyboard and display, to the changing motions and postures.
- Design for a variety of user sizes and user preferences.

Consider that new technologies develop quickly and should be usable at the workstation. For example, radically new keyboards and input devices, including voice recognition, are available and will continue to evolve and improve; display technologies are undergoing rapid changes; wireless transmission no longer limits the placement of display and input devices.

Finally, we can quote directly from a 2008 press release from ANSI, announcing why the standard for sitting upright and related furniture design recommendations were changed:

In an effort to correct the misunderstanding that the 90° posture used in ANSI/HFS 100 (1988) was "the" correct working posture, the furniture chapter now provides four working postures for reference by designers. ANSI/HFES 100 (2007) uses four principal "reference" postures: Standing and three sitting postures: (1) sitting on a reclined seat pan (front edge lower than rear); (2) sitting upright on a horizontal pan; (3) sitting on a declined seat pan (front edge higher than rear) while leaning on a backrest. The standard recognizes that people frequently change their work postures during the day to maintain comfort and productivity. The users' chair should support the three seated postures effortlessly.

ERGONOMIC DESIGN OF THE OFFICE WORKSTATION

Successful ergonomic design of the workstation in the office is a holistic task, because all the interrelated parts of our workstation environment need to be considered (sketched in Figure 5.8) as working together to form the station. Work tasks, work movements, and work activities interact with each other. They affect and are influenced by the workstation components, furniture, and other equipment and by the environment. All of these in turn must "fit" the individual to support his or her well-being and contribute to work output. Of course, job content and demands, control over one's job, and many other social and organizational factors also influence feelings, attitudes, and performance (see Chapter 2).

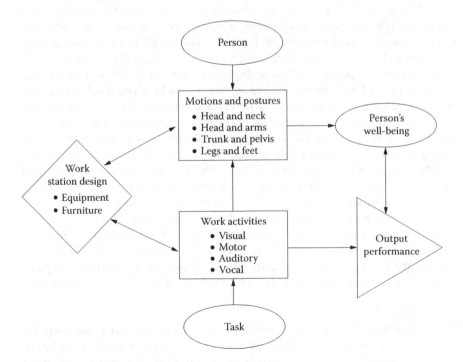

FIGURE 5.8   Person, task, workstation, and performance.

DESIGNING FOR VISION, MANIPULATION, AND BODY SUPPORT

When designing the layout of a work task and workstation, it helps to consider three main "links" between a person and the task:

1. The first link is the *visual interface*: We look at the keyboard, the computer screen, or source documents.
2. The second link is *manipulation*: Our hands operate keys, a mouse, or other utensils; they manipulate pen, paper, telephone, and drawing tools. Occasionally, feet operate controls; starting and stopping a dictation machine is an example. The types and intensities of visual and motor requirements can differ substantially among jobs.
3. The third link is *body support*: the seat pan supports our body at the undersides of buttocks and thighs, and the backrest supports our back. Armrests or a wrist rest may serve as additional support links.

Designing for Vision

The location of the visual targets greatly affects the computer operator's body position. Objects that require our eyesight should be located directly in front of us, at a convenient distance and height from the eyes. If we are forced to tilt or turn our heads up, or sideways, to view our computer screen or document, we are likely to experience neck, shoulder, and back pain, possibly accompanied by eyestrain. It is surprising then that we still often see flawed workstations, with monitor set up too far and high, as in Figure 5.9, and source documents laid flat to the side. In either case, the operator has no choice but to crane his neck, increasing his chance of discomfort or injury.

Neck craning, especially with large displays, may be avoided by using a separate support structure for the monitor so that the display can be adjusted in height, angle, and distance, independently from work surface and keyboard. If the support is spring loaded and moveable by a hand crank or, even better, by an electric motor, adjusting it becomes easy. For traditional desktop computers, the old all-too-common practice of putting the monitor on the CPU box, and possibly also on a stem for angle adjustment, lifts the screen much too high for most users who, as a consequence, tilt their head back and then often suffer from neck and back problems. Instead, the monitor should be located low behind the keyboard so that the user looks down at it. That, fortunately, is a feature inherent to many current computers, especially laptops and touch tablets.

As a rule, the screen or source document should be about half a meter from the eyes, the proper viewing distance for the operator. This is the "reading distance" for which corrective eye lenses are usually ground. A convenient yardstick is to place the screen and source document at arm's length, or slightly less than arm's length.

Document Holder

If you often read from a source document, use a document holder that holds the document close and parallel to the monitor screen, about perpendicular to the line of sight. A document placed far to one side causes a twisted body posture and lateral eye, head, and neck movements—see Figure 5.10.

FIGURE 5.9   The monitor is set up too far and too high.

Corrective Eye Lenses

If you are experiencing eyestrain even though your workstation design is in order, it is time to visit an ophthalmologist or optometrist to have your vision checked. That time usually comes with age: "40 may be the new 30," but not so with vision. The fact is an almost unavoidable consequence of middle age is farsightedness, when natural lenses regularly lose their ability to focus on close visual targets (see Chapter 8), which makes it difficult to discern characters on the computer screen or in a source document. In the United States, this midlife farsightedness is so widely known and accepted that you can buy "reading glasses" in any drug store. If your eye doctor has already corrected your regular distance vision through eyeglasses or contacts, you will be fitted with bifocals or trifocals to correct farsightedness. In these, the lower portion of each lens is ground for reading, which is naturally done by looking downward at the text.

Bifocal or trifocal glasses make it easy to read a document placed on your work surface, but imagine you are reading text on your computer screen, which is placed too high. Wearing bifocals, you would need to throw your head back and tilt your neck at a severe angle to make out the text. The resulting posture is shown in Figure 5.11. Not surprisingly, such extreme kyphosis of the neck can literally cause headaches.

FIGURE 5.10   A document placed far to the side causes a twisted body posture and lateral eye, head, and neck movements. (Modified from a sketch provided courtesy of Herman Miller, Inc., Zeeland, MI.)

Proper Lighting

The computer screen is self-lit, and the office itself should be illuminated at about 200–500 lx (lx, short for lux, is a unit of illuminance, meaning a measure of how much luminous flux—in lumen, lm—is spread over a given area; 1 lx = 1 lumen/m^2). Paper documents may be difficult to read at this fairly low level, so a special task light trained on the document is often helpful, as long as it does not create glare. For offices without computers, the proper illumination range is from 500 to 1000 lx, even higher if there are many dark (light-absorbing) surfaces in the room. Chapter 8 discusses lighting in detail.

Designing for Manipulation

In addition to the eyes, our hands are usually very busy doing various office tasks: grasping and moving papers, taking notes, tapping out phone numbers, and using various computer input devices. If our hands are engaged in many different activities, the varied manipulation is likely to keep our arms and the upper body moving around in our workspace. As a rule with some limitations, the more motion, the better, in contrast to holding still, as in keeping your hand on a mouse.

FIGURE 5.11 Bifocal glasses have their reading chapters at the bottom. Looking up to the display causes severe backward tilt of the head, extreme neck kyphosis, and probably a piercing headache.

When sitting, we do have a large manipulation area available, especially if we move our upper body and, of course, we can cover an even larger area when standing up. However, for finely controlled hand movements, we prefer a space near chest and belly, about 10–40 cm in front of the body—see Figure 5.12. Work done here can also be acutely seen because objects are at a suitable distance from the eyes and so low as to fit the natural downward direction of our gaze.

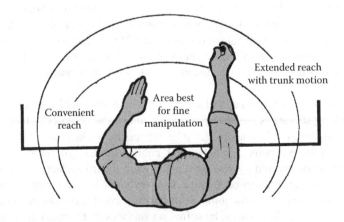

FIGURE 5.12 Areas for gross movements of the hands of a sitting person and for finely controlled manipulations, such as handwriting or keyboarding.

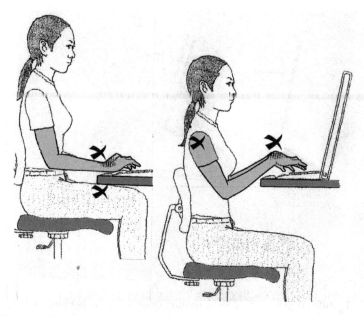

FIGURE 5.13   Keyboard height affects body posture. A keyboard set up too high requires unnecessary muscle tension to lift the shoulders and arm and even to lift and spread the elbows.

Keyboards and other computer input gadgets should all be placed directly in front of our body, at about elbow height, with shoulders relaxed and upper arms hanging at our sides. A keyboard set up higher makes us lift our arms and shoulders, requiring unnecessary muscle tension; setting it too low also creates tension—see Figure 5.13. Placing our keyboard and mouse at different levels, for example, or to the side, or too far away, forces us to move and hold our hands there, which in turn causes muscle tension in our back, neck, and shoulders. This often leads to irritation and pain, even occasionally to repetitive overuse disorders such as bursitis, tendinitis, or cervicobrachial syndrome (see Chapter 7). Chapter 6 contains an extensive discussion of keyboards and other computer input devices.

Resting our wrist or arm on a hard surface or, worse, on a hard edge often occurs when a working surface is pushed up too high, above the operator's elbow height. As Figure 5.14 shows, this leads to sharp local pressure at the point of contact that can cause painful reactions. Examples are cubital tunnel syndrome, where the cubital nerve is compressed by placing the elbow on a hard surface, or carpal tunnel syndrome, where a sharp edge (e.g., of a keyboard housing) presses into the hand's palm area near the wrist joint. Such harmful conditions can be avoided by choosing a low manipulation area, by softening surfaces that support arm and hand (rounding edges, padding surfaces, providing wrist supports) and by adopting better (ergonomically healthier) working habits.

FIGURE 5.14   Arms against a sharp and hard surface.

Designing for Motion and Body Support

We will say it again: our body is built to move around, not to stay still, immobile, over long periods of time. Think about driving for hours on end: you must hold your head in much the same position in order to monitor the road and dashboard instruments and rear-view mirrors, all while keeping hands and foot fixed on wheel and pedal. After a while, it becomes very difficult to stay in the same seated position and posture, even though car manufacturers design car seats as ergonomically as possible. It is really not much different at our computer workstations—where the seat is often much less comfortable than our car seat. When you are driving, it is not always easy or even possible to make a dramatic change in posture, but fortunately, during office work, we usually can get up and move around—and we should.

Since the primary aim of designing an ergonomic workstation is to facilitate body movement, the designer should consider the extreme body postures expected to occur and lay out the interim workspace for motions in between those extremes. In the computerized office, this means to design for walking (standing) and for sitting; the implication for the seat is that it should be designed for relaxed and upright sitting, for leaning backward and forward, and for getting in and out.

Chair design has come a long way and you can find options that have gone beyond the concept of simple user adjustment. They incorporate interesting ideas, like having seat pan and backrest follow the motions of the sitting person and providing support throughout the range, with pan and backrest either moving independently or in a linked manner. Other designs start from the premise that the seat should not be a passive device but an active one: the seat as a whole, or its pan or backrest, can automatically change its configuration slightly over time, perhaps in

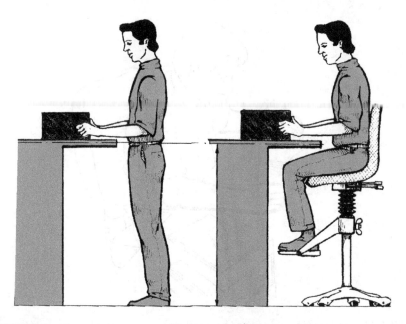

FIGURE 5.15 Encouraging movement.

response to certain sitting postures maintained by the person. The change can be in angles or in stiffness of the material. Seat and back cushions that pulsate were tried in the 1950s to alleviate the strain that military aircrews felt when they had to sit for hours on end to fly extended missions. Should we be reminded by an "intelligent seat" to move, or to get up, after we sat in a static position for some period of time?

In terms of the general office design, in an ideal world we would provide several workspaces for every person. One would be the conventional sit-down desk and computer station. Another would be a stand-up station; then there would be separate areas for meeting with colleagues and visitors, for enjoying a meal or beverage, maybe even to take a nap. In Chapter 4, we discussed how many companies are already embracing mixed-use office plans. Moving around can be encouraged, even designed into the office—see Figure 5.15—by having files stored in a place away from the desk, for example. File cabinets are beginning to vanish from offices— we store files electronically now—so other incentives to keep office employees moving include designing those mixed-use spaces, mentioned earlier.

DESIGNING THE SIT-DOWN WORKSTATION

One of the first steps in designing the office workstation for seated (or standing) work is to establish the main clearance and external dimensions. The size of the furniture should essentially derive from the users' body dimensions and from their work tasks. The main anthropometric data that determine the design, especially the heights, of the furniture that fits sitting users are lower leg (popliteal and knee) heights, thigh thickness, and the heights of elbow, shoulder, and eye.

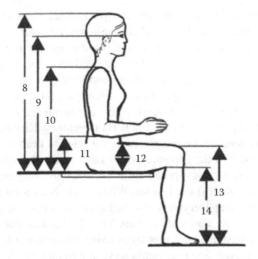

FIGURE 5.16 Body measures commonly taken for sitting individuals. (Adapted from Kroemer et al., 2010.)

For a stand-up workstation, standing eye and elbow heights are important. For more details on anthropometric information and its use for design, see, for instance, publications by Bradmiller (2016), Dainoff (1999), Gordon et al. (2014), Kroemer (2016), and Kroemer et al. (2001, 2010); also check national and professional regulations and standards.

Figure 5.16 depicts body measurements of particular importance to furniture design: # 8 Height Sitting, # 9 Eye Height, # 10 Shoulder (Acromion) Height, # 11 Elbow Height, # 12 Thigh Thickness (Clearance) Height—all measured above the horizontal flat seat pan. Measured from the floor up are # 13 Knee Height and # 14 Popliteal Height. Not shown in this illustration is # 28 Hip Breadth (Sitting). The measuring follows standard anthropometric practice (Gordon et al., 2014; Kroemer et al., 2010). A review of current compilations of anthropometric data demonstrates that such information is available for surprisingly few populations (Kroemer, 2016).

Conventional offices, often inside cubicles, had fixed-height desks and tables set rather high so that people with relatively long legs would fit comfortably underneath. All users had to adjust their postures and seats to those desk and table elevations, which could become awkward, especially if trays or drawers in or on the desk reduced the height of available legroom. Only fairly few offices were equipped with work surfaces, tables, and desks that could be adjusted in height to fit individuals' bodies and work habits.

Nam King's bicycle shop had grown steadily over the last couple of years, so Nam decided it was time to hire office help. His niece Kim had gained work experience for more than a year after finishing trade college and agreed to work for him as sales agent, office assistant, and bookkeeper. They talked about how to set her up in his shop and decided that his mechanic should put a large drawer on the underside of her work table so that, when she wanted to walk away to deal with customer, she could just swipe her laptop computer, work utensils, phone, and purse into it and lock it. That worked fine—but Kim had to set the seat of her office chair on its lowest setting so she could cram her thighs underneath the drawer when she wanted to sit at her table. When sitting to do paperwork or keyboarding, she had to raise her elbows and shoulders and wrists because the work surface was now so high above the seat pan. After just a few weeks, her shoulders began to hurt and several fingers felt numb and tingling at night. She mentioned these problems to her Uncle Nam and they determined that Kim had to improve her posture at work. Since they wanted to keep the drawer, Kim simply moved her seat to the opposite side of table where she adjusted her seat pan at a much higher setting—probably in unconscious compensation for the low seat setting that she had used so far. When Uncle Nam tried to sit there to do some paperwork after everybody else had gone, he found the seat and table uncomfortably high—he was a short man. Luckily, he was also smart, so he decided to buy an adjustable table that, by pressing a button, would adjust higher and lower to fit everybody and even could serve as a stand-up table for him when he grew tired of sitting.

Figure 5.17 illustrates the egregious posture a person would be forced into when no legroom is present underneath the work surface.

Presently, desk-free work is growing ever more popular. Laptop computers, tablets, virtual keyboards, perhaps using voice recognition instead of keying as computer inputs, and software developments together with "wearable" hardware can make solid surfaces of tables and desks unnecessary, cumbersome, and even undesirable.

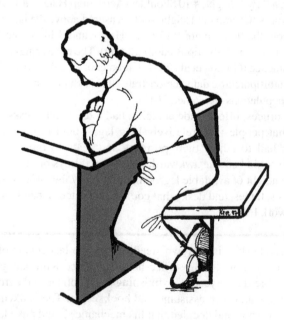

FIGURE 5.17   Sitting with no legroom.

FIGURE 5.18   Niels Different's executive workstation from 1984.

Much of our work can be done on our laps, perhaps with the occasional use of a laptop cart. Niels Different (1928–2013), the famed industrial designer, anticipated that trend in his 1984 "executive workstation"—see Figure 5.18.

Establishing Furniture Heights

There are at least four principal master plans, each with many variances:

1. *Fixed seat height*: The classic design plan, used less frequently of late, starts with a fixed seat height, like we might find in an old-fashioned classroom or in public waiting areas. From there the designer derives the heights of the work surface and, if used, of footrests.
2. *Fixed work surface height*: This traditional furniture design strategy assumes a fixed height of the major work surface, usually the top of table or desk in conventional offices. Next, the designer selects the adjustment range of seat heights that fit the elevation of the work area and from there determines whether footrests are needed for individuals with shorter lower legs.
3. *Fully adjustable furniture*: This modern design procedure assumes that all the furnishings are adjustable in height. The design starts at floor level and from there determines the height adjustments of the seats above the floor. The elevations of seat pans serve to establish the heights of table and of desk surfaces, including the height of supports for keyboard, mouse pad, and other work utensils.
4. *Personal selection*: An individual user chooses and combines particular pieces of furniture to create a special workplace that suits personal tasks and preferences. For example, if the user does "lap work," conventional tables or desks may not be present, possibly replaced with a wheeled tray or laptop cart and mobile supply and storage units. Often a reclining easy chair, possibly with a foot stool, is used instead of a mainstream office seat.

Establishing Depth and Width of Furniture

The clearance depths and widths of the furniture must fit critical horizontal body dimensions of users, especially popliteal and knee depths and hip breadths. The outside depths and widths of the furniture are largely determined by what amount of space the tasks will require and what the users' reach capabilities are.

The common computer workstation has a seat, a support surface for keyboard and display, tablet, or mouse and, mostly for papers, an additional working surface. It is best, and most expensive, to have all of these independently adjustable. Proper workstation dimensions make the office furniture fit almost everybody, tall or short. Suitable approximate height adjustment ranges for office furniture in Europe, North America, and other areas with users of similar body sizes are as follows:

- Seat pan above the floor: 37–51 cm, better to 58 cm
- Support surface for keyboard, mouse: 53–70 cm
- Surface of work surfaces, tables, and desks: 53–72 cm
- Support for the display: 53 to 90+ cm

Work Surface and Keyboard Support

For work surfaces and keyboard placement, there are two main reference points for ergonomic design: the elbow height of the person and the location of the eyes. Both depend on how the workstation user sits or stands, upright or slouched, and how he or she alternates among postures and positions. The table or other work surface of a sit-down workplace should be adjustable in height between about 53 and 70 cm, perhaps even a bit higher for very tall individuals, to permit proper hand/arm and eye locations. Often, a keyboard or other input device is placed on the work surface or connected to it by a tray. A keyboard tray can be useful, especially if the table is a bit high for a person, but it also may reduce the clearance height for knees and upper thighs (think of how the "center drawer" on old desks used to get in the way). A tray should be large enough for keyboard and trackball or mouse pad, unless these are built into the keyboard, but its front edge should not "cut" into palm or forearm.

Keep in mind that specific design recommendations in this book may not apply to user populations with widely different body sizes and work practices who might also prefer different working postures. They would need their own ranges in furniture dimensions and adjustments.

THE OFFICE CHAIR

A chair is a very difficult object. A skyscraper is almost easier. That is why Chippendale is famous.

—Ludwig Mies van der Rohe, famed architect and educator

We have already mentioned in this chapter that the long-held ideal for "proper sitting" at work—upright trunk, with thighs (and forearms) essentially horizontal and lower legs (and upper arms) vertical, all major body joints at zero, 90° or 180°—makes for a convenient but misguiding design template. Such a stance is neither commonly adopted nor subjectively preferred and it is not even especially "healthy." Instead, when designing or selecting an office chair, we must consider a full range of motions and postures.

Seat Pan Design Considerations

When you sit down on a hard flat surface, not leaning against a backrest, you feel the "sitting bones," the ischial tuberosities (the inferior protuberances of the pelvic bones, to be exact) pressing on the seat surface. When we are standing, the gluteus maximus covers those bones, but they are exposed in the seated position. Our upper body's weight is transmitted predominantly by the ischial tuberosities to the seat of the chair, and we feel this especially keenly when the seat pan is hard and unforgiving. The bones act as fulcra around which the pelvic bones rotate under the weight of the upper body. The bones of the pelvic girdle are tightly linked to the lower spine by connective tissue. Rotation of the pelvis therefore affects the posture of the lower spinal column, particularly in the lumbar region. If the pelvis rotation is rearward, the normal lumbar spine lordosis is flattened—see Figure 5.19.

Sitting Comfort and Pressure

To illustrate how pressure and seat pan size affect your sitting comfort in a seated position, let us take an example. Assume your upper body has a *mass* of 50 kg (about 110 lb). If you sit down on a hard surface, the area below your sitting bones transmits all that weight to the seat pan. So, assume that area is 50 cm² (about 20 in²). *Pressure* is defined as *force per area*: the force is nearly 500 newton (N) (50 kg multiplied

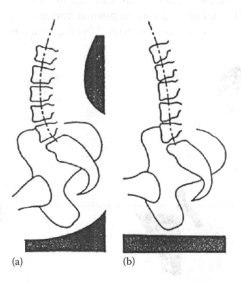

(a)   (b)

FIGURE 5.19   (a) Forward rotation of the pelvic bone on its lowest protrusions (the ischial tuberosities) causes forward bending of the lumbar section of the spinal column (lordosis), which can be enhanced by the forward push of a lumbar pad in the backrest. (b) Rearward rotation of the pelvis can flatten the lumbar spine or even bend it backward (kyphosis).

by 9.81 m/s²). Dividing 500 N by 50 cm² results in 10 N/cm²: this is equivalent to 100 kPa (kilopascal)—about 15 psi—which is the (average) pressure under your ischial tuberosities.

Now, let us give you a larger transmitting surface by making the hard surface soft. This reduces the pressure: increasing the area by the factor 10 reduces the pressure to 1/10th. Consequently, if you temper the seat pan surface with a cushion, upholstery, or elastic mesh, your seated pose just became more comfortable pressure wise.

Strong leg muscles (hamstrings, quadriceps, rectus femoris, sartorius, tensor fasciae latae, psoas major) run from the pelvis area across the hip and knee joints to the lower legs. Therefore, the angles at hip and knee affect the location of the pelvis and hence the curvature of the lumbar spine. With a wide-open hip angle, a forward rotation of the pelvis on the ischial tuberosities is likely, accompanied by lumbar lordosis. These actions on the lumbar spine take place even as associated muscles are relaxed; muscle activities or changes in trunk tilt can counter the effects.

Recognizing this, Staffel (1884) proposed a forward-declining seat surface to open up the hip angle and bring about lordosis in the lumbar area. In the 1960s, a seat pan design with an elevated rear edge became popular in Europe. Since then, Mandal (1975, 1982) and Congleton et al. (1985) again promoted the notion that the whole seat surface should slope fore-downward. To prevent the buttocks from sliding down on the forward-declined seat, the seat surface may be shaped to fit the human underside (Congleton), or one may counteract the downward–forward thrust either by bearing down on the feet (Figure 5.20) or by propping the upper shins on shin pads (Figure 5.21).

A seat pan surface should be tiltable throughout the full range from declined forward, kept flat, to inclined backward, as specified by ANSI/HFES 100 (2007) in the United States. This allows the user to assume various postures with differing curvatures of the lower spinal column, from kyphosis (forward bend) to lordosis

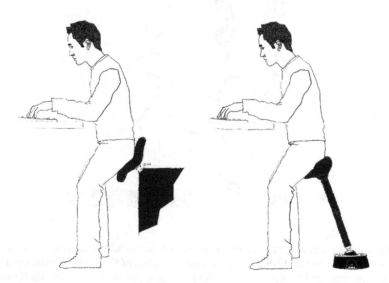

FIGURE 5.20 Semisitting with feet as support.

FIGURE 5.21   Shin pads help to avoid sliding down a semiseat.

(backward bend). The surface of the seat pan must support the weight of the upper body comfortably and securely. Hard surfaces generate pressure points that can be avoided or lessened by cushioning upholstery, pillows, mesh, or other materials that elastically or plastically adjust to body contours.

The only inherent limitation to the size of the seat pan is that it should be sufficiently short that the front edge does not press into the leg's sensitive popliteal tissues, which is located behind the knees. Usually, the seat pan is between 38 and 42 cm deep and at least 45 cm wide. A well-rounded front edge is mandatory. (This is called a "waterfall" in the trade.) The side and rear borders of the seat pan may be slightly higher than its central part.

The height of the seat pan must be widely adjustable, preferably down to about 37 cm and up to 58 cm to accommodate "Western" individuals with either short or long lower legs. It is very important that the person, while seated on the chair, can easily do all adjustments, especially in height and tilt angle. If making adjustments is cumbersome or difficult, they will probably not be made. Figure 5.22 illustrates major dimensions of seat pan and backrest.

Backrest Design Considerations

Two opposing ideas exist regarding backrests: having one versus not having one. Backrest-free proponents embrace the idea of "active sitting," where trunk and leg muscles must remain continually active to keep the upper body in balance—think, for example, of a workstation where the person sits on a flexible ball. This view has few takers; most of us want a backrest. When we lean against it, it carries some of

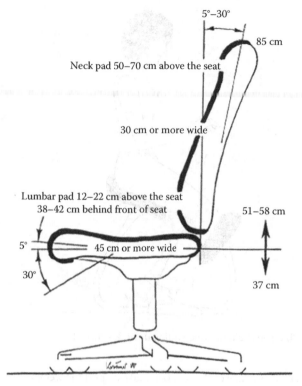

FIGURE 5.22   Major dimensions of the seat.

our upper body's weight and hence reduces the load that the spinal column must otherwise transmit to the seat pan. A second reason is that a lumbar pad, protruding slightly in the lumbar area, helps to maintain lumbar lordosis, believed to be beneficial. A third, related reason is that leaning against a properly formed backrest is relaxing. Of note here is that when you are sitting upright in an office chair, working at a keyboard, you would not be actively using the backrest because to do so would mean pushing against it.

Studies have shown the importance of supporting the back by leaning on a rearward-declined backrest. Andersson et al. summarized the available literature in 1986 and concluded (p. 1113), in some contrast to earlier findings, "in a well-designed chair the disk pressure is lower than when standing." Relaxed leaning on a declined backrest is the least stressful sitting posture. This is often freely chosen by individuals working in the office if there is such a backrest available: "... an impression which many observers have already perceived when visiting offices or workshops with (Display) workstations: Most of the operators do not maintain an upright trunk posture. ... In fact, the great majority of the operators lean backwards even if the chairs are not suitable for such a posture" (Grandjean et al., 1984, pp. 100–101).

Of course, the backrest should be shaped to support the back properly: apparently independently from each other, Ridder (1959) in the United States and Grandjean and his coworkers (1963) in Switzerland found in experiments that their subjects

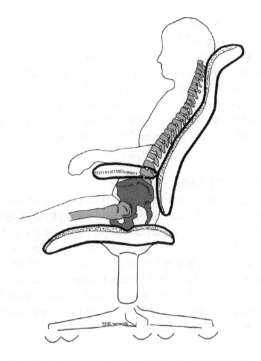

FIGURE 5.23 Preferred backrest and seat pan shape.

preferred similar backrest shapes, as depicted in Figure 5.23. In essence, these shapes follow the curvature of the rear side of the human body. At the bottom, the backrest is concave to provide room for the buttocks; above, it is slightly convex to fill in the lumbar lordosis. Above the lumbar pad, the backrest surface is nearly straight but tilted backward to support the thoracic area; at the top, the backrest is again convex to follow the neck lordosis.

Combined with a properly formed and upholstered seat pan, this shape has been used successfully for seats in automobiles, aircraft, and passenger trains and for easy chairs. The backrest should be as large as can be accommodated at the workplace: this means up to 85 cm high and up 30 cm wide. To provide support from the head and neck on down to the lumbar region, it is usually shaped to follow the contours of our back, specifically in the lumbar and the neck regions. Many users appreciate an adjustable pad or an inflatable cushion for supporting the lumbar lordosis and a similar but smaller pad at the cervical bend. The lumbar pad should be adjustable from about 12 to 22 cm and the cervical pad from 50 to 70 cm above the seat surface.

The angle of the backrest must be easily adjustable while seated. It should range from slightly behind upright (95° from horizontal) to about 30° behind vertical (120°), with further declination for rest and relaxation desirable. Whether or not the seat back angle should be mechanically linked to the seat pan angle appears to be a matter of personal preference.

Armrest Design Considerations

Armrests can provide support for hands, arms, and even portions of the upper trunk; they can help us sit down and get up. Even when only used for short periods of time, armrests can be valuable if they have an appropriate load-bearing surface, ideally padded. They should be adjustable in height, width, and possibly direction. Conversely, sometimes armrests prevent or hinder us in moving our arm, pulling the seat toward the workstation, or getting in and out of the seat. In these cases, having only short armrests, ones that can be flipped to the side, or none, is appropriate.

Footrest Design Considerations

Footstools, hassocks, and ottomans have long been popular to put up one's feet, but footrests in the office usually indicate deficient workplace design—for example, when a seat pan is too high for the individual, perhaps because the seat must be set excessively elevated at a high work surface, then the person uses a footrest. If a footrest is used, it should be so high that the sitting person's thighs are nearly horizontal when the foot is supported. A footrest should not consist of a single bar or other small surface because this restricts the ability to change the posture of the legs. Instead, the footrest should provide a support surface that is about as large as the total legroom available in the normal work position.

The owner of a small suburban consulting firm was perplexed—and, frankly, a little annoyed. He had started his company 3 years ago in the basement of his house and, after 2 years of 14-hour days and tireless effort, had been able to grow his business enough to lease a posh office suite two miles away and hire four associates to help with the work. His most recent employee had joined the company 2 months earlier and her work was excellent. But nonetheless, there was one small aspect about her that chafed. He took great pride in his office and had furnished and appointed it with great care. He was meticulous about choosing not just visually appealing furniture but also items that were properly designed. After all, with a brother-in-law who worked in industrial engineering, he felt he truly understood the importance of well-designed furniture. Accordingly, he had done research, skimped no expense, and purchased the chairs and desks from a reputable office design store. In spite of all of this, however, his newest employee appeared to find her chair less than ideal. Every morning, when she arrived and sat down on her chair, she would place a rolled-up sweater at the small of her back; a few times, she would even set a book along the back of the chair. Last week, she had even brought in a special cushion that she now kept on the lumbar support of her chair. When he asked her why, she indicated with a smile that the chair "just wasn't comfortable" for her. How, he thought, could it not be? The chair had been clearly marketed as "ergonomic" and he had certainly paid extra for this feature. And the other three associates seemed to be perfectly comfortable with their chairs.

Sitting and Back Pain

The posture and movements of the spinal column have been of great concern to physiologists and orthopedists because so many people suffer from annoyance, discomfort, pain, and disorders in the spinal column, particularly in the low back and in the neck areas. Prolonged sitting or standing, preexisting injuries, weak core muscles, lack of exercise, injury during exercise, and intervertebral disk degeneration are all conditions that can lead to back pain. Physical activities and special exercises can improve fitness and back health, but only when done with caution, because some exercises will be contraindicated for those with back issues.

We do best by following the advice to alternate often between stretching, bending, walking, standing, and sitting. If long-time sitting is required, then the design of the seat and other furniture and equipment is especially crucial; for example, a tall and well-shaped backrest that reclines helps to support back and head during work and permits relaxing breaks. While sitting, change position often; you can do this actively by moving in the chair, or (as mentioned earlier) you may be moved passively by pulsating backrests or a slightly shifting seat.

Semisitting

Semisitting is a posture about halfway between sitting and standing, with some of the upper body weight supported at the buttocks and the rest of the weight transmitted through the legs to the ground. Have you ever walked into someone's office and been asked to take a seat on an exercise ball? If so, you have been in a semiseat situation. Semiseats or stand-seats usually do not have full backrests, if any. Figure 5.24 show several examples of semisitting.

Semisitting provides some relief from continuous standing, but is not as supportive as full sitting. Mobility of the trunk is the major advantage of semisitting when there is no backrest. One of the great disadvantages of semisitting is the tendency to slide forward off the support surface. This must be counteracted either by fatiguing

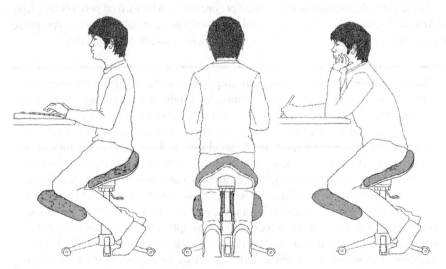

FIGURE 5.24   Examples of semisitting.

leg thrust or by pressure against shin pads, which many people find unpleasant or even painful, though some get used to it. With a low semiseat it can be difficult to move the legs in the confined space between pads and seat as one lowers the body onto the support or arises from it. Some individuals have found semisitting acceptable, and even comfortable, but semisitting should not be generally proscribed because most people prefer more conventional seats.

DESIGNING THE STAND-UP WORKSTATION

Standing up during computer work harkens back to the 1900 office. Moving around and standing, at least for a period of time, is a welcome change from sitting, provided that the person does it at his or her own choice. We might choose to stand for reading, writing, talking with somebody face to face, or telephoning. Stand-up workstations can feature a second computer so work activities can be switched from the sit-down workstation for a while—or, more likely, we simply move a handheld or laptop computer from one workstation to the other. Other stand-up configurations smoothly elevate the work surface so the computer or input device simply shift upward as the person stands up and vice versa when it is time to sit back down. Some people prefer standing and walking altogether to sitting in the office, and some workstations are designed so that computers can be manipulated while the worker moves on a stationary bicycle, stair climber, or treadmill.

Stand-up workstations should be adjustable, with the area used for writing or computer inputs at approximately elbow height when standing, about a meter above the floor. As in a sit-down workstation, the display should be located close to the other visual targets and directly behind the keyboard. If the work surface is used for reading or writing, it may slope down slightly toward the person. A foot bar at about two-thirds knee height (approximately 0.3 m) allows the person to temporarily prop up a foot, which brings out welcome changes in pelvis rotation and spine curvature.

Nonresilient floors, such as those made of concrete, can be hard on people's feet, legs, and backs. Carpets, elastic floor mats, and soft-soled shoes can reduce strain. Appropriate friction between soles and the walkway surface helps to avoid slips and falls.

Cheryl, the manager of a corporate staffing pool, was having difficulty assigning John, one of the administrative assistants. Although John's work was more than satisfactory, his supervisor in the most recent job rotation called him "restless" and said that John's seeming inability to sit still made his colleagues uneasy. Perplexed by this comment—Cheryl had always thought highly of John's work skills—she asked the rotation supervisor to elaborate on John's evaluation. The supervisor explained that John appeared to move around a great deal during his work, even though his duties were largely sedentary in nature. Specifically, most of John's tasks involved using a computer to type in and print out various correspondence and reports. John's output was apparently perfectly acceptable, but he seemed to dislike sitting "normally," as the supervisor put it. Generally, assistants would remain seated at their keyboards for almost their entire 8-hour shift, taking breaks only for lunch or to get coffee. John, however, moved around: he would sit at his workstation sometimes, but would kneel in front of his computer at other times, even work standing up on occasion; additionally, he arose frequently to stretch and take quick strolls. "John does not need a chair," the supervisor commented, "He needs a jungle gym."

DESIGNING THE HOME OFFICE

Nearly everything outlined earlier applies not just to the employer's office but also to the home office. It is a strange phenomenon that we tend to make every effort to design ergonomically at the corporate office, but we may throw out that wisdom when we set up our home work place.

If you do "office work" only for short periods of time in your home, then your dining table and any available kitchen chair may suffice. But as soon as you get serious about using your home office and you are working in it for hours at a time, you should become very conscientious about the working conditions there; Figure 5.25

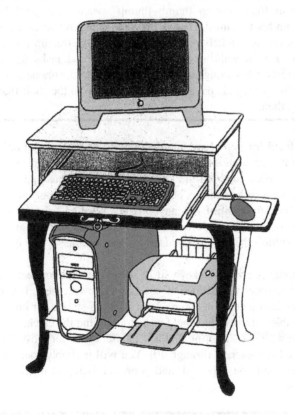

FIGURE 5.25   About everything is wrong in this workstation meant for a home office: the display is too high in the top shelf. The keyboard takes too much space and a sharp edge of the table is likely to cut into the wrists. The mouse pad is too far to the side. The bottom shelf leaves no space for the legs. No comment on the style.

shows a definite no-no. Equip your office with carefully selected furniture, where the components of the workstation fit each other well—and, most importantly, fit you well.

Select an easy chair, a comfortable office chair, a semisit perch, or a kneeling chair—whatever feels good to you and what supports your body well over long periods of time. Perhaps you want to work standing up at least when you read or make phone calls, for example. It is your home office; you can and should set it up in any way you please. And, of course, move around as you see fit. Some of us find as we work at home that we move from one desk to another, from one chair to another, or from one standing station to another; we might pace as we speak on the phone or as we complete a conference call. Simply put, do what feels best for you at any given time.

Lars is a 46-year-old salesman who works half of the time out of his own "office" in a suburb of a large city. He connects with his customers via a telephone with head-mounted speaker and microphone. In the morning, some clients may just barely hear a muffled "thump–thumb–thump" when they speak because he walks briskly on his treadmill. He has mounted a support surface for his note-book computer on the handlebars at the front end of the treadmill so that he can use the computer while walking. For more complex input tasks, he steps off the treadmill and rests in his lounge chair while talking on the phone or working his computer. He has placed his paper files far enough from the chair that he has to get up to grasp them.

Do not fall for furnishings that are unergonomic, such as shelving units that make you put the monitor up high on a rack above the keyboard, or surfaces that do not provide sufficient space for both the keyboard and the mouse pad or other input device. This is your own workspace, put together for your comfort and ease at work, and it does not have to be similar to anybody else's setup—nor does it have to be expensive, because some simple furniture on the market is well designed. Do not fall for a piece of furniture that advertises itself as "ergonomic" if it is not comfortable for you.

This same guidance applies to keyboards; select one that you feel is comfortable (Chapter 6), and consider voice inputs if you can work with them. If you travel with your handheld or laptop computer, consider a docking station for home use, unless you are comfortable with your handheld or laptop around the clock.

Select a room with good lighting that is separate, quiet (for most of us), and well heated and cooled (Chapters 8 through 10). You will probably spend more time in your home office than you expected, and your well-being is worth the effort and money that you spend.

Mike, a young engineering student at a prominent school, Southeastern University, is home for spring break. While he has been away at school for a year, his mother and sister have turned their passion for jams and jellies into a small home-based business. Thanks to a high-quality product, a devoted local following, and recent inroads in distribution through a regional grocery store chain, Marge's Magical Marmalades sales are growing. In fact, Mike's mother (yes, the eponymous Marge) has just acquired two computers for their basement home office; one for her use and one for her daughter's. They plan to use the computers for inventory tracking, logging orders, customer correspondence, billing, and financial reporting.

Of late, Mike's mother has been complaining of back pain and his sister has mentioned aches in her neck. Having just taken a course in ergonomics, he supposes that there may be a connection between their business success and their physical discomfort. He asks for a tour of the basement office. Thrilled about her son's interest, Marge proudly shows him around her hastily refurbished basement, and he slowly looks around and takes in the view. Her beloved new computer is perched on her L-shaped desk, and the monitor is glowing brightly from its position on the long side of the desk. The keyboard rests on the shorter portion of the desk, separate from, and at right angle to the monitor. Marge's chair fits cleanly under the shorter portion of the desk but is too tall to slide under the longer portion; this was why she had placed the keyboard away from the monitor. The chair, unfortunately, does not have height adjustment capabilities. Mike immediately realizes that this configuration forced his mother to twist her body whenever she used her computer, turning one way to access the keyboard, and the other to watch the monitor. "Hmmm," he mumbles. Then, he turns to his sister's workstation. His mother's desk is the only true desk in the cramped office space; since there is no room for another table, his sister's computer is set on top of a three-drawer filing cabinet in the corner. To use it while seated, she can keep the keyboard on her lap, but must place her legs on either side of the cabinet and crane her head upward to view the monitor, which is located about a foot higher than her eyes. "Uhhuhh," he mutters.

Marge turns to her son. "Hmmm?" she says, "uhhuhh? What does that mean?" "Mom," he replies, "I think I know what is up with your back and sis's neck."

WORKING REMOTELY

THE "COFFICE"

Working away from the office is increasingly common, and with the rise of mobile devices it is becoming dramatically easier. Can you think of a recent time when you have walked into a coffee shop and have *not* seen customers poring over sleek laptops as prodigiously as the baristas are pouring coffee drinks behind the counter? (Neither can we). Cafés are convenient workspaces: they offer Wi-Fi, climate control, and refreshments, and they make meeting with other colleagues easy as well. In fact, the term to describe them—coffice—made its way into the Urban Dictionary years ago. However, cafés, bistros, lobbies, waiting areas, airport departure lounges, and hotel rooms are not usually designed for workplace ergonomics even as we attempt to make them our "instant" workstation. You can set up your home office to fit you ergonomically, but not your favorite café, so when you are working remote, you will need to adapt to the environment. Think about your body's biomechanics and now best to support it in a café, and use recommendations already given in this book. We will cover a few guidelines in the following text.

Avoid Sitting over Long Periods of Time

If your café of choice has a bar table, with stools pushed up for seating, claim that spot for yourself, and you will be able to alternate between standing—with the bar stool pushed to the side—and sitting. When you are standing, you can ease your lumbar spine by using the rung of the stool as a foot prop. You can also transform this configuration into a semisitting position by half-leaning, half-sitting on the stool, or staying somewhat tall while supporting yourself with your legs. As you move through these positions at will, assuming you are using a laptop, you will want to tilt the screen forward and backward to easily see the display in accordance with your body's position.

Accessorize Your Laptop

Keep an eye out for devices and add-ons that make your laptop or input device more user-friendly while on the move. Right now, there are several options to make ad hoc workstations more ergonomic; examples include portable laptop stands that raise your laptop screen to a higher eye level, portable mouse track pads (if you use a mouse that needs a track pad), and wireless keyboards. An unattached keyboard is especially useful if you have wrist or hand pain because you can select any keyboard design that works for you, including split key designs.

Use Smartphones Wisely

If you are using your smartphone to work remotely, chances are you are working "on the fly," without the preplanning that goes into packing up your laptop and its accessories. In these cases, use the table or bar that you have selected at the café to prop up your elbows, and then bring the smartphone up to near eye level, holding it with both hands. Use both hands rather than just the dominant thumb to do your keyboarding. You could also take a backpack or folded coat to fashion an off-the-cuff smartphone stand on the table. Use voice dictation when possible to reduce the amount of text you are typing; this, of course, is not always possible in a busy café without irritating your fellow customers. If you are researching topics and using the occasional dictation to search the web, voice recognition can be used without annoying people around you.

Move It!

As always, shift positions when your body tells you to, and stand up for quick breaks when you can. Pay attention to your body's signals and adapt your posture whenever you want.

COWORKING SPACES

Coworking spaces, also called shared offices, are a viable alternative—or supplement—to home offices and corporate offices. The quality of shared spaces varies widely, but there are extraordinary options. Consider working barefoot in Bali, at the Hubud in Ubud, surrounded by bamboo and overlooking rice fields, or taking advantage of the free legal and financial consultations offered by Deskovitz in Amsterdam, or listening to the Start2Bee Band & Orchestra while working at their space in Barcelona, or checking out Third Door in London, where your kids can relax in their registered day care and nursery. (By no means are we endorsing any coworking space in particular; we simply use these examples to illustrate the variety of options based on your coworking needs and preferences.)

Most coworking spaces offer benefits beyond the workstation: refreshments, of course, but also conference room space, scanners and printers, and, in some cases, other unusual amenities like climbing walls and massage spaces or high-end restaurants for meeting over shared meals. They also often provide opportunities for networking, mentoring, collaborating, and learning. The number of coworking spaces on offer has increased dramatically in the last few years, evidence that these places are filling a very real need.

After all, some people find working from home isolating, and cafés or bistros can be distracting and unpredictable (Orozco and Orozco, 2014). Shared workstations alleviate those drawbacks, and they offer what many people look for in their professions: community and collaboration.

As mentioned, many coworking organizations target their appeal to certain business sectors or even niche groups; for example, they may cater to IT start-ups, or to creative writers and artists, or to professionals in the legal industry, or to international businesses. To that end, many offer special services geared toward the needs of that particular group, like translation services, photo studios, design labs, incubators, and accelerators.

There is some irony here, and it is not lost on us: for many decades, we as a collective workforce have decried our plight at the office—the annoying commutes, the pointless office politics, the drudgery of shared spaces, and the irritations of forced togetherness—and now, many of us seek out and even pay for the privilege of sharing workspaces.

WORKSPACE OF THE FUTURE

Time will tell what tomorrow's office will look like, but—as in the past—it will be shaped in large part by the prevailing advances in technology. More of us than ever are working remotely; if this trend continues, workspace options will evolve to reflect that. If there are enough of us "road warriors" and digital nomads, then suppliers of potential "landing" places will want to figure out what we are looking for and what we are willing to pay extra for. As an example, large café chains are presently human engineered for efficient café operations rather than for remote workers' comfort and efficiency. Will they adapt and offer some "ergonomic workstations" in the near future? Perhaps charge for the space and provide free beverages? Add some private space for meeting rooms?

A FEW MORE WORDS ABOUT FURNITURE ADJUSTMENTS

Traditionally, experts have emphasized the need for easy adjustments of chairs, of support surfaces for the keyboard, and of the display in height, distance, and angle—and we have repeated these sentiments in our earlier discussion. At the same time, many formal studies (e.g., by Vitalis et al., 2000) supported our personal experiences that current adjustment features are seldom used. We conclude that existing means of adjustment are still too complicated. Think of getting into a car's passenger seat, knowing that you will be in it for a while, and trying to adjust it. How long does it sometimes take to locate the appropriate levers and buttons? Another related inference is that adjustments should be automatic and not require any conscious interaction of the individual. For example, ideally, the seat should simply follow all movements of the sitting person, "reading" the body, and supporting it throughout. We are hopeful that engineers will soon use technological progress so that furniture will automatically adjust itself to our proportions as we shift and move.

FITTING IT ALL TOGETHER

An uncomfortable office is not instantly converted into a fantastic one by changing out a computer or by simply installing a better chair. Sure, tweaking any one component will help, but all of the components, equipment, and furniture—see Figure 5.26—plus lighting and climate must fit each other, and the person in the office must be willing and able to take advantage of all the offered possibilities. It is that holistic approach we mentioned at the beginning of this chapter; each component of the office interacts with the other elements.

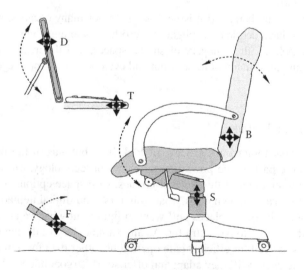

FIGURE 5.26   All components of the workstation must fit each other: Seat pan S and backrest B, keyboard support T and display location D, and a Footrest F may be needed.

Once you have designed or updated your corporate or home office, be prepared for some real strides in your attitude toward your work, because improving your office environment often brings about positive change in mental attitude. Imagine that you are running a division of a large company and you are in charge of their newest branch. You are given the go-ahead and the budget to design the office, and you have done your research about what employees seem to like and what current trends are in office design. After you have solicited employees' input and done your due diligence as far as research, you complete the office design. The new design features natural light, well-monitored temperature, open workspaces with some collaborative spaces and other amenities, and ergonomic workstations. This shows your staff that you value them and you care about their well-being. Knowing that you have invested in them, that you want them to be happy and healthy, in turn is mirrored in how your employees feel about jobs. This new, "healthy" office encourages movement, gets people to leave their chairs to communicate and collaborate, and creates an environment where people are stimulated, where they thrive, where they want to succeed, and which they will not want to leave.

Put differently, the workplace has a profound effect on individuals' well-being, and the office is the manifestation or representation of the company itself. Should not a responsible, caring employer want to create a healthy, lively, and enjoyable environment for their employees? Showing people how important they are results in less illness, better performance, higher retention, and, at the end, better profitability.

ERGONOMIC DESIGN RECOMMENDATIONS

Neither theories nor practical experiences endorse the idea of one single proper, healthy, and comfortable sitting position. Instead, many motions and postures are subjectively comfortable for one particular person, at that particular time, depending on body, preferences, and work activities.

The traditional postulate that everybody should sit upright is debunked and furniture should be designed for free-flowing motion. Changing from one posture to another one, moving freely among all the comfortable poses, is ideal. Motion, change, variation, and adjustment at an individual's will—this is central to well-being.

Consequently, furniture should allow for body movements among various postures. For this to happen, furniture should adjust automatically or at minimum must be easily adjustable in all of its primary dimensions, especially seat height, seat pan angle, and backrest position. The entire computer workstation should facilitate easy variations as well, especially location and height of the input devices and height and distance of the display.

ERGONOMIC GUIDELINES

We should redesign our computer systems and how we interact with them to better fit our abilities—see Chapter 6 of this book. But even if we do not fundamentally change our office tools and habits, we still can make our work as easy as possible by doing the following:

Place all the things you must operate with your hands (keyboard, mouse, trackball, pen, paper, telephone)

- Directly in front of you
- At elbow height
- Within easy reach

Place all the things you must clearly see (display, source document, writing pad, template, keyboard—see also Chapter 6)

- Directly in front of you, at your best viewing/reading distance (currently, your computer display is probably too far away from you)
- Low behind your keyboard

Sit on a seat designed so that you can change your posture frequently. If long-term sitting is required, then a tall backrest that can recline helps support back and head.

Change your body position often. Change helps avoid continued compression of tissues, especially of the spinal column, facilitates blood circulation, and counteracts muscular fatigue—and it breaks the boredom.

Support your arms and hands by resting them, as often as feasible, on soft arm rests attached to the seat or on padded wrist rests at the keyboard—best use both—but avoid hard surfaces and especially rigid corners and edges that compress the skin tissues.

Keep the shoulders relaxed, the upper arms hanging down, the forearms horizontal, and the wrists straight. To achieve this, all elements of your workstation (chair, computer on its support, table and desk) must be carefully arranged in concert with each other.

ERGONOMIC TROUBLESHOOTING

My eyes are tired, teary, and achy!

- Place all the things you must clearly see (display, source document, writing pad, template, keyboard)
  - Directly in front to the best viewing/reading distance
  - Low behind your keyboard, avoiding the use of a tall tilt stand under the monitor
- Be sure there is no light—from a window or from a task lamp—reflected in the display or shining directly into your eyes (see Chapter 7)
- If the condition persists, have your eyes checked by a specialist (ophthalmologist or optometrist)
- Talk to your supervisor and get a medical evaluation if the condition does not abate.

My back hurts!

- Place all the things you must clearly see (display, source document, writing pad, template, keyboard)
  - Directly in front to the best viewing/reading distance
  - Low behind your keyboard, avoiding the use of a tall tilt stand under the monitor
- Place anything you must operate with your hands (keyboard, mouse, telephone, paper)
  - Directly in front of you
  - At elbow height
  - Within easy reaching distance
- Take a break at least every 30 minutes; stand up, move around, and take a quick walk.
- Make sure you can lean comfortably against the backrest of your seat.
- If your chair is simply not comfortable for you, find one that more closely matches your sitting habits.
- Talk to your supervisor and get a medical evaluation if the condition does not abate.

My neck hurts!

- Place all the things you must clearly see (display, source document, writing pad, template, keyboard)
  - Directly in front to the best viewing/reading distance
  - Low behind your keyboard, avoiding the use of a tall tilt stand under the monitor
- Place anything you must operate with your hands (keyboard, mouse, telephone, paper)
  - Directly in front of you
  - At elbow height
  - Within easy reaching distance
- Take a break at least every 30 minutes; stand up, move around, and take a quick walk.
- Make sure you can lean comfortably against the backrest of your seat.
- If your chair is simply not comfortable for you, find one that more closely matches your sitting habits.
- Talk to your supervisor and get a medical evaluation if the condition does not abate.

My shoulder hurts!

- Take a break at least every 30 minutes; stand up, move around, and take a quick walk; do some shoulder stretches.
- Put your mouse or trackball next to the keyboard, at elbow height.

- Operate the mouse or trackball with your other hand—it will feel odd at first, but you'll get accustomed to it.
- Use your armrest and wrist rest often.
- Are you frequently on the phone? You may be cradling the phone on your shoulder, be sure to request/obtain a headset instead.
- Talk to your supervisor and get a medical evaluation if the condition does not abate.

My wrist/hand hurts!

- Take a break at least every 30 minutes; stand up, move around, and take a quick walk.
- Make sure that your wrist remains straight while working the keyboard or other input device.
- Strike the keyboard keys lightly.
- Use armrest and wrist rest often.
- Put the mouse or trackball next to the keyboard.
- Talk to your supervisor and get a medical evaluation if the condition does not abate.

My leg hurts!

- Take a break at least every 30 minutes; stand up, move around, and take a quick walk.
- Make sure that you have ample room at your workstation to move and reposition your legs and feet freely and often.
- If the front portion of your seat presses on the underside of your thighs
  - Lower the seat (most likely, you also must lower keyboard and monitor, table, or desk accordingly)
  - Change out your chair to one that has a soft "waterfall" shape on front of the seat
  - Use a wide and deep footrest
- Talk to your supervisor and get a medical evaluation if the condition does not abate.

REFERENCES

ANSI/HFES 100. (2007). American National Standards Institute (ANSI) (1988), *American National Standard for Human Factors Engineering of Video Display Terminal Workstations* (ANSI/HFES Standard No. 100-1988.) Santa Monica, CA: Human Factors and Ergonomics Society.

ANSI/HFS. (2008). Sitting pretty: New American national standard addresses workstation and computer design. [Press Release].

ANSI/HFS 100. (1988). *American National Standard for Human Factors Engineering of Visual Display Terminal Workstations.* Santa Monica, CA: Human Factors Society.

Bendix, T., Poulsen, V., Klausen, K., and Jesnen, C. V. (1996). What does a backrest actually do to the lumbar spine? *Ergonomics* 39, 533–542.

Bonne, A. J. (1969). On the shape of the human vertebral column. *Acta Orthopaedica Belgica* 35(Fasc 3–4), 567–583.

Booth-Jones, A. D., Lemasters, G. K., Succop, P., Atterbury, M. R., and Bhattacharya, A. (1998). Reliability of Questionnaire Information Measuring Musculoskeletal Symptoms and Work Histories. *American Industrial Hygiene Association Journal* 59, 20–24.

Bradford, E. H. and Lovett, R. W. (1899). *A Treatise on Orthopedic Surgery* (2nd ed.). New York: William Wood & Company.

Bradford, P. and Byrne, W. (1978). *Chair.* New York: Crowell.

Bradtmiller, B. (2016). Anthropometry in human systems integration. Chapter in Boehm-Davis.

Congleton, J. J., Ayoub, M. M., and Smith, J. L. (1985). The design and evaluation of the neutral posture chair for surgeons. *Human Factors* 27, 589–600.

Corlett, E. N. and Bishop, R. P. (1976). A technique for assessing postural discomfort. *Ergonomics* 19, 175–182.

Dainoff, M. J. (1999). Chapter 97: Ergonomics of seating and chairs. In W. Karwowski and W. S. Marras (Eds.). *The Occupational Ergonomics Handbook.* Boca Raton, FL: CRC Press, pp. 1761–1778.

Dickinson, C. E., Campion, K., Foster, A. F., Newman, S. J., O'Rourke, A. M. T., and Thomas, P. G. (1992). Questionnaire development: An examination of the Nordic musculoskeletal questionnaire. *Applied Ergonomics* 23, 197–201.

Gordon, C., Blackwell, C. L., Bradtmiller, B., Parham, J. L., Barrientos, P., Paquette, S. P. et al. (2014). 2012 anthropometric survey of U.S. army personnel: Methods and summary statistics. Technical report NATICK/TR-15/007. U.S. Army Natick Soldier Research, Development and Engineering Center, Natick, MA.

Grandjean, E. (1963). *Physiological Design of Work* (in German). Thun, Switzerland: Ott.

Grandjean, E. (1987). *Ergonomics in Computerized Offices.* London, U.K.: Taylor & Francis.

Grandjean, E., Huenting, W., and Nishiyama, K. (1984). Preferred VDT workstation settings, body postures and physical impairments. *Applied Ergonomics* 15, 99–104.

Helander, M. G. and Zhang, L. (1997). Field studies of comfort and discomfort in sitting. *Ergonomics* 40, 895–915.

Kaufman, L. (2014). Google got it wrong. The open office trend is destroying the workplace. *The Washington Post.* Https://www.washingtonpost.com/posteverything/wp/2014/12/30/google-got-it-wrong-the-open-office-trend-is-destroying-the-workplace/.

Kroemer, K. H. E. (2016). *Fitting the Human* (7th ed.). Boca Raton, FL: CRC Press.

Kroemer, K. H. E., Kroemer, H. B., and Kroemer-Elbert, K. E. (2001). *Ergonomics: How to Design for Ease and Efficiency* (2nd ed.). Upper Saddle River, NJ: Prentice Hall/Pearson.

Kroemer, K. H. E., Kroemer, H. J., and Kroemer-Elbert, K. E. (2010). *Engineering Physiology* (4th ed.). Heidelberg, DE: Springer.

Kuorinka, I., Jonsson, B., Kilbom, A., Vinterberg, H., Biering-Sorensen, F., Andersson, G., and Jorgensen, K. (1987). Standardized Nordic questionnaires for the analysis of musculoskeletal symptoms. *Applied Ergonomics* 18, 233–237.

Lehmann, G. (1962). *Praktische Arbeitsphysiologie* (2nd ed.). Stuttgart, Germany: Thieme.

Mandal, A. C. (1975). Work-chair with tilting seat. *Lancet* 1, 642–643.

Mandal, A. C. (1982). The correct height of school furniture. *Human Factors* 24, 257–269.

MedicineNet. Definition of neutral posture. Published online June 2012. Http://www.medicinenet.com/script/main/art.asp?articlekey=25488. Retrieved October 2015.

Merrill, B. A. (1995). Contributions of poor movement strategies to CTD solutions or faulty movements in the human house. In *Proceedings of the Silicon Valley Ergonomics Conference and Exposition, ErgoCon 95.* San Jose, CA: San Jose State University, pp. 222–228.

Ridder, C. A. (1959). *Basic Design Measurements for Sitting* (Bulletin 616, Agricultural Experiment Station). Fayetteville, AR: University of Arkansas.

Shackel, B., Chidsey, K. D., and Shipley, P. (1969). The assessment of chair comfort. *Ergonomics* 12, 269–306.

Staffel, F. (1884). On the hygiene of sitting (in German). *Zbl. Allgemeine Gesundheitspflege* 3, 403–421.

Tenner, E. (1997a). How the chair conquered the world. *Wilson Quarterly* 21, 64–70.

Tenner, E. (1997b). The life of chairs. *Harvard Magazine*, January–February, pp. 47–53.

van der Grinten, M. P. and Smitt, P. (1992). Development of a practical method for measuring body part discomfort. In S. Kumar (Ed.). *Advances in Industrial Ergonomics and Safety IV*. London, U.K.: Taylor & Francis, pp. 311–318.

Wright, W. C. (1993). *Diseases of Workers, Translation of Bernadino Ramazzini's 1713 De Morbis Articum*. Thunder Bay, Ontario, Canada: OH&S Press.

Zacharkow, D. (1988). *Posture: Sitting, Standing, Chair Design and Exercise*. Springfield, IL: Thomas.

6 Keyboarding and Other Manual Tasks

I already have it, but a good keyboard is invaluable when you spend a lot of time typing. My favorite one is the ancient IBM Model M I have at home.

—Markus Persson, a.k.a. Notch, programmer and founder of Mojang

Never trust a computer you can't throw out the window.

—Steve Wozniak

A computer once beat me at chess, but it was no match for me at kickboxing.

—Emo Phillips, entertainer and comedian

OVERVIEW

Long-ago offices were known as "white-collar factories," and they were considered a far safer alternative for the labor force at that time than working in manufacturing or as a laborer. After all, agrarian work involved using dangerous tools; the shipping industry entailed heavy lifting; railroad labor was rough and dirty; all are more treacherous than putting pen to paper. In the early days of office work, writing words and numbers by hand was the most common task—and one that carried with it a surprising risk of injury. Maintaining a firm grasp on a pen and guiding it ceaselessly in finely controlled motions over paper generates a repetitive strain of the musculoskeletal system that can cause repetitive strain injuries (RSIs). They called it writer's cramp 300 years ago, and now we call it RSI; terminology may have changed, but it is as prevalent today as it was then.

Of course, modern-day "writer's cramp" is now usually a result of keyboarding rather than pen-to-paper work, and keyboarding creates a whole new host of ergonomic tasks and challenges.

THE HUMAN HAND: WHAT IT CAN AND CANNOT DO

The hand is a remarkably versatile part of our body. It can touch and grip, manipulate forcefully and with fine control, hold delicately, and move energetically. It is enduring and sturdy, but fallible—it can be injured by a sudden blow or cut. More relevantly to us in the office, it can also be damaged by often-repeated small impacts that may come from external vibrations or from our own voluntary internal efforts, like pressing keys on a keyboard.

Some of us are more susceptible to such repetitive motion overload, and some of our jobs require more repetitive motions per day than others. Biomechanically, our

121

hands are designed to do many different activities, interspersed with breaks for resting or doing other tasks, but they are not equipped to do the same action over and over for hours on end, inherent in an 1850s scrivener's job of continuous hand writing or in today's seemingly endless tapping on keyboards.

For our own comfort and well-being, we should vary our office tasks over the course of a day, shake up our surroundings and movements, and use our manifold mental and manual capabilities. Keyboarding all day, for example, can be both boring mentally and stressful physically. So, let us take a look at our office equipment, especially keyboards, and redesign or rearrange them from an ergonomic perspective, while also reevaluating our other office tasks to fully utilize our capabilities without overtaxing them.

In this chapter, we examine the biomechanics of hands and wrists to help us understand what they can do and what their limitations are; we explain the evolution and layout of keyboards, what they can do, and what their limitations are; and we discuss how the two interact and interrelate. Since this interaction between our bodies and our offices can lead to RSIs, we also cover this type of injury, and we present guidelines and ergonomic design recommendations on how to safely work on office devices like keyboards while minimizing risks of injury.

Note: You may skip the following part and go directly to the "Ergonomic Design Recommendations" section at the end of this chapter—or you can get detailed background information by reading the following text.

MANUAL OFFICE TASKS

HAND AND WRIST

The bones and joints of the hand and wrist give our body the strength and flexibility we need to manipulate objects in a myriad of ways. Each hand contains 27 bones: 8 compact bones next to the forearm, 5 long ones leading to the knuckles, and 14 slim bones inside the thumb and the fingers. The bones can move against each other in gliding joints, kept in place by tough elastic tissues, known as *ligaments*. The pull of muscles generates voluntary movements. Some muscles are located within the hand (they are called intrinsic muscles), but the major strong muscles are inside the forearm (termed extrinsic muscles). *Tendons* are cable-like tissues, similar to ligaments, except that they attach on one end to bone and, on the other end, to muscle (as opposed to bone-to-bone attachment of ligaments). Contraction of a muscle pulls on its tendon, which transmits that displacement to the bone. The tendon slides inside a *sheath* that provides guidance and lubrication.

As shown in Figure 6.1, eight small bones, called "carpals," are located adjacent to the radiocarpal articulation, the wrist joint. They are firmly bound by ligaments in two rows of four bones each, forming an elastic block called the "carpus." On its proximal side, the carpus connects with the ends of the *ulna* and *radius*, the bones of the forearm; on the distal side, it connects with the *metacarpals*. On the palmar (inner) side of the hand, the carpals form a canal-like structure that is covered by a tough ligament; the resulting opening is called the "carpal tunnel."

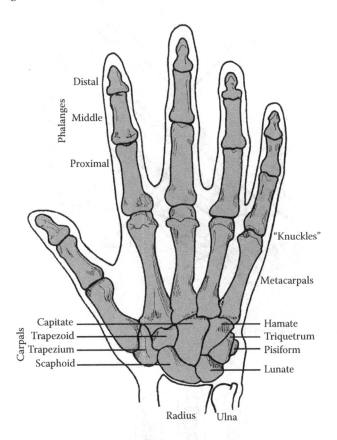

FIGURE 6.1   The bones and ligaments of the hand.

Ligaments and nerves extend through the carpal tunnel to reach the palm.

Five long metacarpal bones are kept attached to the carpus by ligaments; on their distal ends, the metacarpals provide the basic joints (the "knuckles") for the five digits of the hand. The distal head of each metacarpal rounds to form a joint with a *phalanx*, one of the bones inside the fingers. These oval joints allow for circular motion of the fingers at their bases. Each of our four fingers contains three phalanges, connected by hinge joints between each other; our thumb has only two phalanges. The phalanges that connect with the metacarpals are called "proximal phalanges," those at the ends of the digits are the "distal phalanges," and there is a *middle* one in each of the four fingers.

Figure 6.2 sketches the back of the hand with its tendons; muscle pulls on these tendons extend (straighten) the five digits, four fingers, and the thumb. The illustration shows that the cable-like tendons are, over much of their lengths, encapsulated in sheaths. Figure 6.3 shows, with a bit more detail, a palmar view of the sheaths encapsulating the tendons that bend (flex) the hand digits.

The flexor tendons coming from the muscles in the forearm must, after crossing the wrist joint, then pass through the tight passage of the carpal tunnel; as mentioned earlier, the carpal tunnel is formed by carpal bones and the

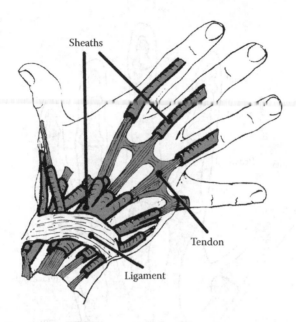

FIGURE 6.2   The back of the hand.

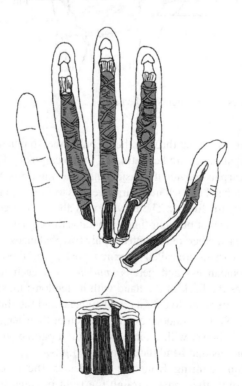

FIGURE 6.3   Palmar view of the sheaths encapsulating the tendons.

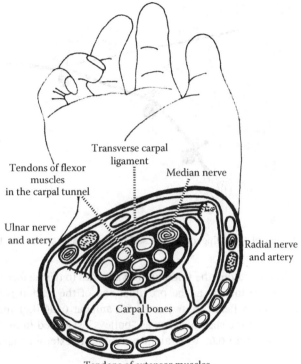

FIGURE 6.4   Flexor tendons and the median nerve pass through the carpal tunnel.

transverse carpal ligament. Figure 6.4 shows schematically how blood vessels, the median nerve, and flexor tendons thread through the narrow opening of the carpal tunnel. (When looking at the backside of your hand, the dorsal side, you can easily see the extensor tendons on that side of the carpals and metacarpals.) Increased pressure inside the carpal tunnel, often caused by swelling of tissues due to inflammation, compresses the enclosed blood vessels, nerves, and sheathed tendons.

Figure 6.5 provides a side view of the flexor tendon of a straight finger. Tough ringlike and crossed sections of ligaments are parts of the tube-shaped sheath that

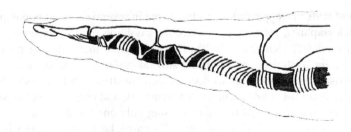

FIGURE 6.5   Side view of the flexor tendon of a straight finger.

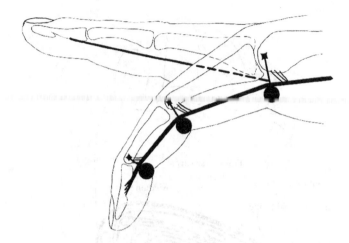

FIGURE 6.6   Side view of a bent finger showing the "pulley" arrangement of the flexor tendon.

keeps the tendon in place along the finger until it attaches to the surface of the distal phalanx. Shortening the muscle at the proximal end of the tendon pulls it inside the sheath until the finger flexes in its joints with annular (ringlike) and cruciform (cross- or x-like) parts of the sheath acting as "pulleys," sketched in the simple biomechanical model of Figure 6.6. To straighten a digit, its extensor muscle pulls on the extensor tendon.

The movements of tendons can be very large—9 cm or more as Treaster and Marras (2000) showed—which means that tendons must slide that far within their sheaths. Tension force within a tendon can be multiples of the force that is actually applied with a fingertip. When a joint is angled, the tendons crossing it must be bent accordingly; the bending force causes pressure on the sheathed tendon, which adds additional strain on the tissues of tendons and sheaths.

Repetitive Strain from Keying

Now that we understand how our hands and wrists work, it is not so surprising that frequent and repetitive movement, especially around contorted joints, can lead to wear and tear of the tendons and their sheaths and, consequently, to RSI. Any highly repetitive activity can be punishing and may overexert the human musculoskeletal system.

Reports of RSIs first appeared in the 1800s and in several different populations—in Morse telegraphists, for example, where the disabling injury was simply called "telegraphist's wrist" (Kroemer, 2001). They were also found among musicians, keyboardists, and violinists, who did and still do suffer from aches and pains in the hand, wrist, and forearm regions due to repetitive stress. Their medical diagnoses were frequently listed as tendinitis, tenosynovitis, and carpal tunnel syndrome. Some well-known pianists had to resort to using only one nonafflicted hand while playing; others had to stop playing altogether. Extensive keyboard operation has long been known to be mentally and physically stressful. For much more information on when RSIs and cumulative stress disorders first were researched and explained, please see Appendix C, where we provide a detailed overview of both early and more recent literature covering these topics.

In 1926, Klockenberg reported heartbreaking stories of typists: lured by available jobs in "secure" offices, many young women trained to become typists. After just a few years working on then purely mechanical typewriters, some began to suffer from numb, painful hands, with injuries debilitating enough that they could not lift their small children. Figure 6.7 is an illustration of the working conditions at that time. In the early 1960s, when the coauthor of this book was on his first job, he was directed to research the working conditions that led to health problems in the hands and arms of operators of "electric" typewriters.

Evidently, overuse disorders began occurring from the first use of typewriters, as the literature from the early 1900s indicates. Heidner (1915) and Klockenberg (1926) attempted to alleviate these injuries by designing new keyboards that minimized repetitive motion. In 1951, Lundervold published the first report of his groundbreaking electromyographic experiments to attain knowledge about the use of individual muscles while typewriting. He investigated 135 typists, including 88 patients, most suffering from "occupation myalgia." Lundervold's studies finally provided clinical evidence supporting the long-held opinion that repetitive typing could and did lead to a cumulative overexertion injury. Lundervold's findings established that such ailments as tendinitis, tenosynovitis, and tendovaginitis of the upper extremities were indeed "occupational diseases" of typists.

FIGURE 6.7 Typical posture of the typists in the 1920s. This image is also shown in Figure 5.1 to show the stooped posture.

Since then, waves of health complaints have been reported among keyboard users: first in the 1960s and 1970s in Japan, then in the early 1980s in Australia, followed by outbreaks among newspaper reporters and other keyboarders in the United States in the 1990s (Kroemer, 2001). Some of these may have had social and psychosomatic aspects but real ailments were often diagnosed. As Kroemer et al. (2001) reported, the underlying biomechanical and pathological events are well understood.

Repetition-related injury often occurs in the carpal tunnel, with syndromes related to increased pressure in the carpal tunnel, reduced blood supply, and reduced functioning of the median nerve. (See Chapter 7 for more details on this topic.) Overexertion near the wrist, and especially of the tendon/sheath unit, is fundamentally a mechanical overuse problem. Think of the tendon as a cable under tension that is rubbing vigorously against the inner surface of its sleeve (the tendon sheath) with lubrication (by synovial fluid) that fails when it is overused. The conditions worsen if tension and transmitted force increase, if the use frequency grows, and if the tendon is bent due to a flexed, extended, or laterally deviated body joint. Biomechanically, similar conditions can occur in any number of work sites—on the shop floor in manufacture or assembly, for example, or on the construction site with carpentering or bricklaying, and in the office while keying or using a computer mouse. In many workplaces, there is a large and pressing need for better design of manual tasks and, relatedly, of equipment; this is especially true for computer data input.

Before we move on to equipment design, we would like to note that those individuals prone to these types of injuries—and that is most of us—can be proactive in protecting ourselves from them. If we are aware of our own working habits, we can learn to tap keys more gently, investigate alternative keyboard designs, keep our shoulders and arms relaxed, use padded wrist supports when helpful, keep our wrists straight (see Figure 6.8), and—particularly—take breaks, as many as needed, to give our bodies a break.

FIGURE 6.8   Elevating or lowering the wrist impedes the smooth functioning of the tendons that move thumb and fingers. Note that the posture problems shown here with "old" computers are the same that exist with new equipment, as depicted in Figure 5.13.

KEYBOARD DESIGN

Origin of the QWERTY Keyboard

Throughout the nineteenth century, many inventors tried to replace manual penmanship with mechanisms that could print on paper, by "typewriting machines." They commonly followed the traditional practice of composing text from single letters, numerals, signs, and spaces. Their efforts were primarily directed at designing and constructing machinery that could generate the imprint as a consequence of an operator pushing a button ("key"). Given the technology at hand at that time, this proved to be a formidable task, and few of the proposed mechanisms were workable; into the 1870s, none was commercially successful.

The various typewriting machines featured a multitude of keyboards: some similar to the long black and white keys on the piano, some featuring double or triple rows of button-like keys, and some showing keys arranged in concave or convex circle segments (Adler, 1997; Herkimer County Historical Society, 1923; Martin, 1949). While they were busy constructing the actual typing mechanisms, the would-be inventors did not worry much about how to assign letters, numbers, or other signs to specific keys, nor were they concerned with how to arrange them for easy use. "Human engineering" was not part of the design task in the nineteenth century.

C. Latham Sholes was the first inventor to successfully design, produce, and market a typewriter, and with it came a special keyboard. His typewriting machine, patented on August 27, 1878 (U.S. patent 207,559), shows a keyboard with four rows of altogether 44 keys—see Figure 6.9. The keys on the third row are labeled,

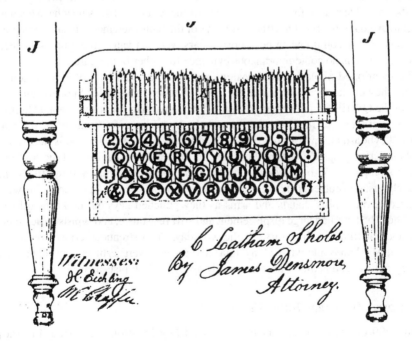

FIGURE 6.9   Sholes' keyboard with 44 keys.

from the left, QWERTY. This arrangement of those six keys is often used nowadays as a short label for the arrangement of all the letter ("alpha" or "alphabetic")
keys. Remarkably, the current English-language QWERTY keyboard is still essentially true to Sholes' original design, with only the positions of the letters C and X
exchanged from the original and the letter M moved one row.

Sholes had previously obtained a series of patents on typing machines (in 1868,
79,265 and 79,868; in 1876, 182,511; and in 1878, 199,382, 200,351, 207,557, and
207,558), several of which were coauthored with other inventors. His first 1868 patent
(79,265) showed two rows of keys, "similar to the key-board of a piano," as he wrote.
His next 1868 patent (79,868) also had 2 rows of keys, with 13 keys each, but now all
in the same flat plane. The keys alternate in length and are lettered in numerical and
alphabetic order; Sholes did not give an explanation for this key layout. His 1876 patent and his first three patents of 1878 all exhibit similar arrangements, each with three
rows of button-like keys affixed to lever-type bars. Those buttons show no assign-
ments to letters or numbers.

Why Sholes decided on any of these key arrangements is unknown. In his final patent
207,559, Sholes made 14 specific technical claims, but none refers to the key selection or
layout. One drawing in this patent shows a frontal view of the invention with four stag-
gered, horizontal rows of keys, with the row farthest from the operator the highest. Another
drawing depicts a top view of the 4 straight rows, each with 11 keys, shown in Figure 6.9.

Apparently, the 1878 QWERTY layout contains some remnants of an alphabetic
arrangement. Sholes was a printer by trade, so we can surmise that the placement
showed similarities to the arrangement of the printer's "type case," in which pieces
were assorted according to convenience of use and not according to the alphabet.
Another possible reason for the arrangement of the letters and keys may have been an
effort to avoid having typebars (the early keys of the then-used mechanical typewriter)
stick together or collide when neighboring bars were activated in a quick sequence.
However, there is no contemporaneous evidence for either of these guesses.

We mentioned Sholes' varied designs and patents—and lack of explanation as
to why they came about—to make the point that the reasons for Sholes' layout of
keys in the QWERTY pattern are obscure. Certainly, Mr. Sholes was not a "human
factors engineer". In fact, whether or not the layout is even suitable for the way
we use keyboards today or how typists used typewriters decades ago is debatable.

Sholes' typographic machine became the predominant device to type text and
numbers on paper. With millions of mechanical typewriters employed in offices and
privately, usage problems inherent to the design soon became apparent. The work
required of the typists' hands and arms as they punched keys on the keyboard was
tiresome. The keys had large displacement and stiff resistance, and typists struck them
thousands of times in the course of the workday. Such strenuous effort overloaded
many typists' hands and wrists: "myalgia" and related overexertions, reported earlier
in telegraphists and pianists, now frequently appeared in typists as well.

Ergonomic Keyboard: The 1915 Version

From 1909 on, several patents for improved key locations appeared, but they
kept to Sholes' original layout with its bent columns and straight rows. In 1915,

Heidner advocated substantial changes in the basic keyboard layout. He obtained U.S. patent 1,138,474; on its first page, he wrote that he had "invented certain new and useful improvements" in keyboards "to enable the operator to obtain a better view of the keys and to write with greater ease, in a less cramped position than ordinarily. With this object in view, I divide the keyboard into halves and locate the two groups of keys thus formed in such manner that the forearms of the operator ... instead of converging, lie substantially parallel with each other. ... [T]he hands have not to be twisted outward ... and there being thus much less strain upon the abducent muscles, writing is rendered considerably less fatiguing."

In addition to splitting the keyboard into left and right halves and placing them at a slant angle to allow better forearm posture, Heidner also arranged the keys in curved rows "in accordance with the natural form of the hand, that is to say, lengths of the fingers. ... [C]onvergence of the key groups further facilitates operation of the keys by the fingers in their natural position in the extended axis of the forearm." Figure 6.10 is taken from his 1915 patent and shows his design recommendations.

However, no typewriter manufacturer adopted Heidner's improvements; in fact, his recommendations were seemingly forgotten since they were not even mentioned in the literature or in subsequent related patents for about 70 years. Heidner's astute observations preceded scientific research by at least a decade—such as that published by Schroetter (1925) and Klockenberg (1926). This research measured effort and recorded fatigue associated with typing. Proposals similar to Heidner's work reappeared later (without referring to him); examples include propositions in the 1940s by Griffith, in the 1960s by Kroemer, and thereafter by many other inventors and manufacturers of improved keyboards in the last decades of the twentieth century. Figure 6.11 shows a keyboard that is split into halves for each hand. The sections can be tilted down to the sides. The keys are arranged to follow the natural motion paths of the hands' digits.

The mechanical nature of the early typewriters made it impractical to change the basic lever system and hence the design of the keyboard. Around 1950, electric auxiliary power reduced the amount of energy that the operator had to apply in each keystroke—although this did not diminish the frequency of keystrokes. Then, around 1960, electronics began to replace mechanics. The new technology would have allowed new and ergonomic designs of keyboards, but even in 1988 the ANSI/HFS 100 (1988) explicitly proscribed a conventional QWERTY keyboard. In the 1980s and 1990s, more keys were added on the sides and on top of Sholes' design so the total number of keys easily doubled. Standardization perpetuated the convention of keys arranged in straight horizontal rows with zigzag columns on the QWERTY part but in straight columns on all other key sets.

PROBLEMS WITH CURRENT "CONVENTIONAL" KEYBOARDS

Certainly, the most serious problems with the use of keyboards stem from the excessive motions of the wrist and hand digits to accomplish the needed key activations. Let us take some hypothetical examples of keyboard users and demonstrate the

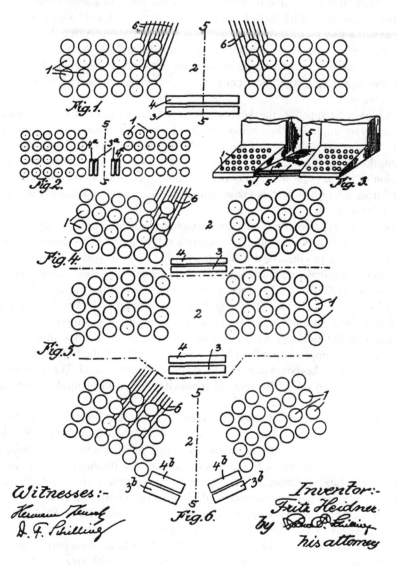

F. HEIDNER.
TYPE WRITING MACHINE.
APPLICATION FILED MAR. 19, 1914.

1,138,474.

Patented May 4, 1915.

FIGURE 6.10 Heidner's keyboard, 1915.

enormity of what they ask their hands and wrists to do every day (yes, we advocate for taking frequent breaks throughout this book, but let us postulate for ease of calculation that no breaks are taken by the hypothetical people in this paragraph).

Many hands simply cannot take such frequent mechanical strain, and they respond with irritation and possibly inflammation.

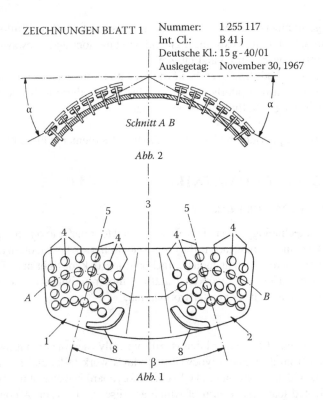

ZEICHNUNGEN BLATT 1   Nummer: 1 255 117
                       Int. Cl.:   B 41 j
                       Deutsche Kl.: 15 g - 40/01
                       Auslegetag: November 30, 1967

Schnitt A B

Abb. 2

Abb. 1

FIGURE 6.11   Split-half keyboard.

As a part-time college student, Joan is a relatively slow keyboarder, typing only 30 words a minute, but to finish her 4-hours-per-day online school work, her fingers complete 36,000 digit bends and 36,000 digit lifts a day. Susan is a photojournalist who types at a speed of 50 words a minute (with one word containing 5 letters) for 6 hours a day to document the stories about the photos she takes. This means she bends her fingers 90,000 times to press down keys, followed by the same number of finger elevations to release the keys. As a writer, Ben types 60 words for 8 hours a day, meaning 144,000 key presses and 144,000 return movements for his fingers.

Other problems that commonly arise are also related to design details of the keyboard. They include

- Zigzag key columns on the QWERTY key pad, but straight columns on other keypads—hard on fingers and mind
- Straight rows of keys, which are not aligned to fingertips—resulting in strain and fatigue
- Horizontal rows of keys that enforce extreme inward rotation (pronation) of the forearms—strain and fatigue

- Large numbers of keys in some keyboards (up to 130 on "full-size" keyboards) that require extreme digit and wrist motions and sideways stretch of the little finger—strain and fatigue
- Arms and hands held over the keyboard with no support—strain and fatigue
- QWERTY layout poorly suited for tablets and other touch screen devices when typing with thumbs—slow output

More details are provided by Kroemer (2010) and Kroemer et al. (2001).

ERGONOMICS OF DATA ENTRY

DESIGNING THE MOTOR INTERFACE

With current technology, most of the data we need to transfer to our computer is done by hand; the common interface is the conventional flat keyboard, often accompanied by other input means such as mouse, trackball, joystick, or light pen. The design of these interfaces affects the workload and the motor activities of the operator and dictates the layout of the computer workstation.

Keyboard

As we have noted, the keyboard that is still typically used in computing today is the one derived from a long-ago typewriter. Today's work tasks and technology raise reservations with the conventional QWERTY keyboard because of its ergonomically contraindicated features, several of which are discussed earlier. A consequence of the widespread use of this keyboard is overuse disorders, which are common in keyboard operators. Causal or contributing factors are the frequency of key operation combined with awkward forearm and wrist postures.

Keyboard Modifications

Many proposed improvements to the QWERTY keyboard have been suggested, and we can categorize those into several broad categories. Some relocate letters on the keyboard and change the geometries of the keyboard, such as arranging keys in curved rows and columns. Others divide the keyboard into halves, one for each hand, arranged so that the center sections are higher than the outsides, so hands do not need to pronate as required on the flat keyboard. Still others involve activating two or more keys simultaneously—called chording—to generate one character or whole words, or chunks of words, such as those used by court reporters. And still others recognize that keys do not have to be of the conventional binary (on/off) tap-down type but may be toggled or turned and have three or more different contact positions.

To illustrate, Figure 6.12 shows an example of an "ergonomic" keyboard, the ternary chord keyboard (Langley, 1988, U.S. patent 4,775,255), which has only four keys for each hand. It allows the hand to rest on built-in wrist pads, and the keys can support the fingers because the keys are toggled rather than tapped down—see Figure 6.13. Learning to operate such an unusual input device proved to be surprisingly easy and fast (McMulkin and Kroemer, 1994).

FIGURE 6.12   The ternary chord keyboard.

FIGURE 6.13   Keys are toggled rather than tapped.

Taking a closer look at these alternatives, we can further categorize them into the following types (adapted from Cornell University Ergonomics Web):

1. *Modified standard layout*: This keyboard looks like a standard keyboard except that the keys are angled so that there is less ulnar deviation when typing.
2. *Fixed-angle split keyboards*: These keyboard designs split the alphanumeric keys at a fixed angle and they slightly tent the keyboard. There is some evidence of reduced discomfort because of minimized ulnar deviation (lateral bending of the hands). These designs work better for broader, larger frame individuals or pregnant women because they put the arms in a better position to reach around the front of the body.

3. *Adjustable-angle split keyboards*: These keyboard designs allow the user to change the split angle to suit their own needs; the user decides on the split angle, but this means that they will need some training.
4. *Completely split keyboards*: In these designs, the left-hand and right-hand portions of the keyboard are completely split apart. In some designs, the keys are presented in a scooped design that allows the hands to rest in a more neutral posture for typing.
5. *Vertically split keyboard*: The design is like that of an accordion and the user types with the hands facing each other; consequently, the keys cannot easily be seen. This design works well to reduce ulnar deviation and wrist extension, but it is important not to have the keyboard too high; otherwise, the chest and shoulders can fatigue.
6. *Chord keyboards*: These have a smaller number of keys and letters, and digits are generated by combinations of keys in chords. The learning curve is quite steep, but these are extremely useful to users with special needs (i.e., arthritic hands).
7. *Specialist keyboards*: Several different keyboard designs have been developed to assist users who have some physical limitation or who wish to type in a different way. One device allows the user to rest his or her hands on a series of switches that detect different directions of finger movements, and these generate the characters. Another device lets users rest their hands on two domed surfaces and then move these surfaces to generate the characters.
8. *One-handed keyboards*: These keyboards are for users with extreme physical limitations, or where one hand needs to key while the other performs another action; the Half-QWERTY uses the same kinds of keys that are found on a regular keyboard, but each key functions in two modes to generate all of the characters of a regular keyboard in a smaller area.

In the past, new designs have been rejected with the argument that it would be too difficult and time-consuming to learn working on a different key set—even if increases in typing speed and accuracy could ultimately be achieved. (Examples are alphabetic keyboards and Dvorak's and Langley's designs.) Given how prevalent repetitive overexertions are, it is questionable whether we should even want increased keying throughput. There is much anecdotal experience—plus the experimental evidence with the Langley design (mentioned earlier)—that indicates that we can learn to operate newfangled input devices with surprising ease and speed. Who would have predicted around 1990 that, within two decades, mini keyboards, physical or virtual, would become the de facto standard on millions of electronic phones?

Many of the key arrangements on cellular phones are, surprisingly, close to the traditional QWERTY; however, the tiny size of the keys disregards specs in ANSI/HFS 100 (1988, 2007) and earlier test results regarding suitable spatial location, appropriate dimensions of the overall board and specifically of each key, and appropriate key traits in displacement and resistance. Obviously, key users are willing and able to accept new designs and can become proficient with them.

We know that people vary in size, shape, ability, and preference, so there is no one keyboard that is best for every person. The conventional large QWERTY keyboard is likely to remain in common office use because it is mass produced at low cost and it is entrenched. People are familiar with it; however, it is not the best design for all individuals and all situations. The varied ergonomic designs mentioned earlier have been developed to address specific needs under specific work conditions, and some are viable alternatives for QWERTY. Of course, given that most of us are accustomed to the traditional keyboard, we should expect to experience a learning curve as we are acquiring new skills on new key sets. What is more, with new technologies, sensors, and activation devices, we expect that there will be several different ways to transfer input signals to computers other than keyboards. Some QWERTY-less means already exist, including voice recognition devices; we will visit these in the text below.

Other Input Devices

In addition to the traditional key (discussed earlier), a variety of other input devices can be used, including

- *Mouse*: A palm-sized, hand-contoured block with one or more finger-operated button(s) commonly slid on a surface (mouse pad), mostly used to move a cursor
- *Puck*: Similar in shape to a mouse but typically has a reticular window used on a digitizing surface (tablet)
- *Trackball*: A ball mounted in an enclosure whose protruding surface is moved by palm or hand digits, usually to move a cursor
- *Joystick*: A short lever, operated by a fingertip or, if larger, by the hand, typically for moving a cursor for pointing or tracking
- *Stylus*: A pencil-shaped, handheld device often used for object selection, freehand drawing, and cursor movement, usually on a tabletop digitizing surface
- *Light pen*: Similar to a stylus but commonly used on a cathode ray tube display surface
- *Tablet*: A flat, slate-like panel over which a stylus or puck or just the tip of a finger is moved, usually for cursor movement and object selection
- *Overlay*: An opaque overlay of a tablet that provides graphics
- *Touch screen* (*touch sensitive panel*): An empty frame, or overlay, mounted over the display screen that locates the position of a finger or pointing device, used for object selection, object movement, or drawing

Typical uses of these devices are to move a cursor, input a single character or number, manipulate the screen content, digitize information, and point to insert or retrieve information, but the tasks change with evolving technology and software.

The design of the keyboard or any other manipulated input device, especially of a mouse or puck, determines hand and arm posture. The essentially horizontal surface of conventional keyboards and of most mice requires that the palm also be kept approximately horizontal; this is uncomfortable because it requires strong pronation in the forearm, close to the anatomically possible extreme. A more convenient angle of forearm and wrist rotation is achieved by a sideways-down tilt of the mouse surface that is in contact with the palm. This is similar to splitting a keyboard and tilting the sections down to the side.

Hands and wrists are not the only body parts affected by the designs of keys and other input devices. The location of hand-operated input devices of any kind naturally determines the position of the hand, of the forearm, and of the upper arm and, consequently, even affects the posture of the trunk. As a rule, the best position for the hand is in front of the body, at elbow height. If mouse, puck, stick, or other input instruments are used jointly with a keyboard, they should all be placed closely together; in fact, many keyboard housings already contain a trackball or trackpad.

Supporting arm and hand appropriately relieves shoulder and back muscles from their holding tasks. Various kinds of arm, wrist, and hand rests can be employed during work, or at least during breaks in extended work. Such rest periods are helpful but by themselves are not sufficient to overcome unsuitable equipment and excessive workloads.

New Solutions

Traditionally, we input data into our computers through manual interaction between our fingers and such devices as keys, mouse, trackball, or light pen. These entry techniques force us into a complex task sequence:

- *First, breakdown*: We must break down formulas, sentences, and words into letters, numerals, and symbols.
- *Second, association*: We must make an association between each letter, numeral, or symbol to a distinct key, another device, or a function element.
- *Third, separate entries*: We must manually operate keys and the like for separate entry of the numerals, symbols, and letters into the computer.

After all of this, we then use the computer program to reconstruct the original words, sentences, and formulas, which we had just broken into components. Whew! What a time-consuming, complicated use of the human mind, and what a lot to ask in terms of intensive manual work. In a sense, we are wasting efforts of both our bodies and minds.

It would be much better to transmit the message from the human to the computer as a whole, at least in batches. For this, voice communication with the computer is an obvious and (now) technically feasible solution. Speech recognition converts speech into machine-readable text, meaning a string of character codes. This technology appeared on the verge of breaking through years ago but then dead-ended in seemingly unsuccessful attempts to implement it; it appeared to haphazardly work but had many flaws. Now, the stage is set for speech recognition: processors are incredibly fast and strong, storage—with the debut of the cloud—is virtually unlimited, Wi-Fi is everywhere, mobile apps are multiplying, and new algorithms and data about speech have been learned and incorporated. In certain functions, speech recognition is already widely used in concert with a keypad, such as in transcription of medical or legal dictation, journalism, and even writing essays or novels. Speech becomes an especially interesting alternative when we want to input data hands-free, perhaps while walking or otherwise physically taking a break.

We especially like the idea of using voice recognition as a supplement to keyboard input on a regular basis. Using a combination of voice and keyboard operations, and maybe throwing a mouse or stylus into the mix, could be an extremely efficient and effective way of entering information into a computer, speeding up work performance while resting our hands and wrists, and reducing our chances of a musculoskeletal injury.

Using our imagination, it seems clear that many other ways exist to generate computer inputs. Consider that we might use

- Hands and fingers for pointing, gestures, sign language, and tapping
- Arms for gestures, making signs, and moving or pressing control devices
- Feet for motions and gestures and for moving and pressing devices
- Legs for gestures and moving and pressing devices
- Torso, including the shoulders, for positioning and pressing
- Head, also for positioning and pressing
- Mouth for lip movements, tongue, or breathing such as through a blow/suck tube
- Face for grimaces and other facial expressions
- Eyes for tracking

Combinations and interactions of these different input signals could conceivably be used similarly to the way in which we utilize them during face-to-face communication. Of course, the techniques developed to recognize these signals must be able to clearly distinguish them from environmental clutter or other "loose energy" that could interfere with sensor pickup.

Such ideas are not far-fetched, neither in sensor technology nor in human input activities. Even if we return to our keyboard, and if it is the old QWERTY, but we do so in concert with the use of voice commands, all while resting our wrists on appropriate padding during frequent breaks, our computer input could be far more efficient. Moreover, we could alleviate many of the cumulative trauma disorders that keyboard users endure.

In the United States, ANSI/HFS standards and HFES guidelines refer to specific design features such as key spacing, key size, key actuation force, and key displacement. They also provide design dimensions for mouse and puck input devices, trackballs, joysticks, styli and light pens, tablets and overlays, and touch sensitive panels. Keep in mind, however, that the ANSI standard merely specifies "acceptable" applications based on accepted human factors engineering research and experience and that the standard does not apply to operator health considerations or work practices. The standard permits alternate computer workstation technologies in order not to impede development and use of novel solutions.

ERGONOMIC RECOMMENDATIONS

Ergonomic Keyboards

Consider using a modern "ergonomic keyboard," for example, one that

- Has a split and slanted QWERTY (alpha) key section
- Has a split QWERTY section whose halves are tilted down to the sides
- Has a "key feel" that you like (mostly in terms of key resistance and travel)

- Has built-in cursor controls (such as trackpad, trackball, nipple) that you find easy to operate
- Is designed and built to the newest technology and human engineering standards
- Has only as many keys as you need

CONVENTIONAL KEYBOARDS

If you continue to use a conventional keyboard, select one that

- Has a "key feel" that you like (mostly in terms of key resistance and travel)
- Has built-in cursor controls (such as trackpad and trackball) that you find easy to operate
- Is designed and built to the newest technology and human engineering standards and guidelines
- Has only as many keys as you need

SPECIAL KEY SETS

If you need a special key set occasionally, such as a numeric pad, consider getting one in addition to your regular keyboard and placing that special keypad carefully in front of you; do not "compromise" between them.

TRY NEW KEYBOARDS

Try out as many keyboards as you can. To really evaluate them, you may have to use them for hours or even days. Do not hesitate to discard a keyboard if you do not like it; keyboards cost much less than treating repetitive strains does. Remember that the variety of keyboards available is extensive, and you should use the one that you and your body prefer.

OTHER INPUT DEVICES

Consider voice communication with your computer. Evaluate carefully whether to use a mouse, trackball, or trackpad for cursor control, puck, joystick, stylus, light pen, tablet, or touch screen for pointing, tracking, digitizing, drawing, and other special tasks.

FINALLY...

Once you have selected the input device that you really like, sit in a truly comfortable chair (see Chapter 5) and place the manual input device

- Directly in front of you (with the display closely behind it) over your thighs
- At or below elbow height (with your upper arms hanging from your relaxed shoulders)

You may consider working standing up, at least once in a while. For this, you need an easily adjustable workstation or a second computer set up for standing operation. Many of us already use a laptop or handheld computer; those, of course, allow for a great deal of freedom in moving your workstation around.

If it is helpful, use a padded wrist rest (or armrest) at least during inputting breaks while you remain at your workstation.

Take short breaks from keyboarding or other manual inputting often. It is best to leave your workstation completely for a while.

REFERENCES

Adler, M. (1997). *Antique Typewriters. From Creed to QWERTY.* Atglen, PA: Schiffer.

ANSI/HFS 100. (1988). *American National Standard for Human Factors Engineering of Visual Display Terminal Workstations.* Santa Monica, CA: Human Factors Society.

Greenstein, J. S. (1997). Chapter 55: Pointing devices. In M. Helander, T. K. Landauer, and P. Prabhu (Eds.). *Handbook of Human–Computer Interaction* (2nd ed.). Amsterdam, the Netherlands: Elsevier, pp. 1317–1348.

Heidner, F. (1915). Type-writing machine. Letter's Patent 1,138,474, dated May 4, 1915; application filed March 18, 1914. United States Patent Office, Alexandria, VA.

Herkimer County Historical Society (Ed.). (1923). The story of the typewriter 1873–1923. Published in *Commemoration of the 50th Anniversary of the Invention of the Writing Machine.* Herkimer, NY: Author.

Klockenberg, E. A. (1926). *Rationalization of the Typewriter and of Its Use* (in German). Berlin, Germany: Springer Verlag.

Kroemer, K. H. E. (2001) Keyboards and keying: An annotated bibliography of the literature from 1878 to 1999. *International Journal Universal Access in the Information Society.* 1(2), 99–160. www.springerlink.com/index/yp9u5phcqpyg2k4b.pdf.

Kroemer, K. H. E. (2010). 40 years of human engineering the keyboard. In *Proceedings of the 54th Annual Meeting of the Human Factors and Ergonomics Society.* Santa Monica, CA: Human Factors and Ergonomics Society, pp. 1134–1138.

Kroemer, K. H. E., Kroemer, H. B., and Kroemer-Elbert, K. E. (2001). *Ergonomics: How to Design for Ease and Efficiency* (2nd ed.). Upper Saddle River, NJ: Prentice Hall/Pearson.

Langley, L. W. (1988). Ternary chord-type keyboard. Patent 4,775,255. United States Patent Office, Alexandria, VA.

Martin, E. (1949). *Die Schreibmaschine und ihre Entwicklungsgeschichte (The Typewriter and Its Development al History).* Aachen, Germany: Basten.

McMulkin, M. L. and Kroemer, K. H. E. (1994). Usability of a one-hand ternary chord keyboard. *Applied Ergonomics* 25, 177–181.

Schroetter, H. (1925). Knowledge of the energy consumption of typewriting (in German). *Pflueger's Archiv fuer die Gesamte Physiologie des Menschen und der Tiere* 207(4), 323–342.

Treaster, D. E. and Marras, W. S. (2000). A biomechanical assessment of alternate keyboards using tendon travel. In *Proceedings of the XIVth Triennial Congress of the International Ergonomics Association and 44th Annual Meeting of the Human Factors and Ergonomics Society.* Santa Monica, CA: Human Factors and Ergonomics Society, pp. 6-685–6-688.

7 Safety at Work

I see you wear safety goggles on your forehead. I also like to live dangerously…

—Austin Powers: International Man of Mystery

I will say that I cannot imagine any condition which could cause a ship to founder. I cannot conceive of any vital disaster happening to this vessel. Modern shipbuilding has gone beyond that.

—Captain EJ Smith, captain of the Titanic

OVERVIEW

In the ideal world, we would skip off to our offices in the morning, delighted to launch into another day of fulfilling, well-paying, lauded, and interesting work. In Chapter 2, we covered a number of theories of job satisfaction, where the ultimate goal of self-actualization is on the top of the ladder or pyramid. But let us remember that there are a number of levels below that top echelon. Performing well depends on more than employee skills and corporate reward incentives. So before we get to the "skipping to work" stage, we must turn our attention to workplace design once again.

From the "hierarchy of needs" model we learn that basic needs like food and shelter must be met before we can even think about meeting higher-order needs. In the workplace, these basic needs include safety and security. You probably do not routinely ask yourself these questions when you step into your office in the morning, but is your company doing whatever it can to keep your work environment free of unnecessary hazards? Are you concerned about an outbreak of fire? Is the work you do secure, or is it subject to theft or plagiarism? Is your Internet connection really private? Do cables and electrical cords get in your way? Are you asked to work night shifts? How much do you use the telephone? Does your neck hurt, or your wrist?

This chapter covers safety and security issues at work, whether at a corporate office, a home office, or any other remote work location. We also discuss the specific challenges faced by those employees who work uncommon work shifts, like overnights.

FIRE SAFETY

Fire is a threat to the people in the office, to their personal belongings and their work, as well as to equipment and the building. Fire safety is a high-priority demand; not only does heat and fire kill but so do toxic fumes and smoke. The U.S. Bureau of Labor Statistics reported 143 workers died from fires in 2011. On an average day, more than 200 fires occur in U.S. workplaces, according to the Occupational Safety and Health Administration (OSHA). According to OSHA, more than 5000 are injured annually in explosions and fires on the job, and the annual cost of workplace fires to American businesses is more than $2 billion (U.S. Department of Labor). In the United States, employers are required to comply with specific fire safety and health standards, which are issued and enforced by OSHA or by one of the 27 OSHA-approved State Plans. When a state plan applies to a given organization, it must be at least as effective as OSHA's.

Ergonomic office design includes the notion that, for the protection of office occupants and their property, fire emergency responders must be able to respond, enter, locate the emergent incident, and resolve it most quickly. Avoiding fire hazards is a special technology, and one we will leave to the experts to expound upon. We can, however, offer a few guidelines related to offices:

- Power cords should be inspected regularly for wear and be replaced if they are frayed or have exposed wire; also, cords should never be used if the third prong has been damaged or removed.
- Cords should never overload outlets; the most common cause of fires started by extension cords is improper use and overloading. If you are using extension cords, ensure that they are approved by a certifying laboratory and use them only temporarily.
- If employees use space heaters or if you are using a space heater in a home office, make sure that the appliances have a switch that automatically shuts them off if they tip over. Space heaters should never be placed near combustible materials like paper and should not be left unattended if you leave your office or home.
- If your office has fire sprinklers installed, never place objects too closely below fire sprinkler heads to allow a full range of coverage.
- You must know how to use a fire extinguisher. Whether you are working at home or in the office, you should know where the fire extinguisher is and ensure that it is in good working order.

Alerting people in the office to an existing combustion is another ergonomic challenge. The first task is to alert people of the existence of a danger, and the second task is to guide them away from it. As the specialized literature describes in much detail, a combination of acoustic and visual alarm signals works most reliably to alert people of danger. Usually, a loud (high intensity) signal is used that pierces the regular noise level through carefully selected sounds (in terms of frequencies and intensities), best modulated over time. Ideally, the sound signal is accompanied by flashing red lights. Typical everyday examples are the sirens and strobe lights used on police and firefighter vehicles.

In office buildings, escape paths must be kept open and distinctly marked, with indicators clearly visible even in smoke, in the dark (electricity may be out), and in crowded conditions and perceptible by people in a state of panic. There must be at least two means of escape, remote from each other, and well marked. The escape routes must also be usable by disabled individuals, such as those in wheelchairs. Importantly, fire doors should never be blocked or locked; this applies not only to the inside but also to the outside. More often than you would think, a back door—one of the fire exits—is treated as an afterthought and may be blocked on the outside by garbage bins or parked cars. Employers must ensure that these egresses are clear and unobstructed. If the space is a high hazard area, exit doors need to swing in the direction of exit travel.

Emergency lighting is necessary; again, remember that there may be no electricity, so emergency lighting should function independent of electricity. Fire drills are a must, even if they appear to be a nuisance. Ideally, there should be a written emergency action plan for evacuating employees, and it should describe the routes to use and the procedures to follow.

INDOOR AIR QUALITY

We spend most of our time indoors, working in an office, working at home or remotely, or simply going about our daily routine. Whether you work primarily behind a desk or in an industrial setting, indoor air quality can be a real, serious occupational health hazard—and one that is often overlooked. Many pollutants go unnoticed because they appear harmless, but the secondhand tobacco smoke, the carpet in a home office, the potted plant in the corner of a corporate office, and even your colleague's strong perfume can be harmful to certain individuals.

For most of us, the main contributors to poor air quality include inadequate ventilation systems, water damage and mold growth, office overcrowding, cubicle design that blocks off air flow, too much or too little humidity, cleaning chemicals and pesticides that are stored incorrectly, and poor housekeeping with dust and dirt remaining in our work environment.

Indoor air quality issues are especially dangerous for those who suffer from asthma—in the United States, roughly 1 in 12 adults, close to 19 million people, have this condition (CDC, 2015). Dust that accumulates around cluttered areas, strong odors (regardless of the source), and mold commonly found in the soil of potted plants can be annoying for all of us and trigger severe attacks for those who have asthma. Newer buildings can be worrisome for people with asthma if they feature heavily insulated doors and windows installed for energy savings. These insulation technologies make modern buildings nearly airtight, which saves on climate control costs but reduces air flow. This lack of airflow can result in polluted indoor air with high levels of carbon dioxide and water vapor, both of which appeal to dust mites and encourage their breeding. The presence of nuisance pests, including dust mites, mice, and cockroaches, is noxious to asthma sufferers and causes severe attacks (CDC, 2015).

Asthma takes a significant toll on employee productivity. In the United States, it is responsible for 14 million missed workdays (CDC, 2015). Asthma is also a leading reason for hospitalizations and trips to the emergency room, so ensuring that your workplace prioritizes lung health for all makes good financial sense (American Lung Association, 2015).

Fortunately, there are some basic guidelines we can follow to improve overall air quality:

- Keep the environment smoke-free. Even in countries where smoking is banned in public and most other buildings, like in the United States, many people are exposed to secondhand smoke during their workday. There is *no* safe level of secondhand smoke. And if you are working out of a remote location or home office where smoking is present, do your best to eliminate it, prohibit smoking indoors, and ask smokers to partake outside, well away from doors and windows.

- Establish fragrance-free policies; this can be a sensitive topic, because some people cannot resist splashes of aftershave and spritzes of perfume. The fact is, though, "one person's perfume is another person's poison" because there are people with heightened sensitivity to fragrance, and even more people simply do not enjoy the scent that another person might like. A unified fragrance-free policy serves to encompass all workers and is becoming more and more common. Wearing perfumes, colognes, aftershave, scented lotions, or hairsprays are all disallowed under these policies, and this avoids any ambiguity.

- Maintain the office or home heating/ventilation/air-conditioning (HVAC) system and ensure it is functioning well. If you are working at home, set the temperature to suit yourself; if you are in an office, choose a temperature that meets most employees' preferences. The American Society of Heating, Refrigerating, and Air-Conditioning Engineers (ASHRAE) suggests selecting a temperature that is agreeable to at least 80% of individuals, with the range of 21°C–23°C (69°F–73°F) a general recommendation. In summertime, when outdoor temperatures are higher, consider keeping air-conditioned offices slightly warmer than during colder seasons to minimize the temperature discrepancy between indoors and outdoors.

- Take flu shots—offered free, we hope: prevention is the best medicine when it comes to influenza.

- Keep pests at bay, preferably through integrated pest management (IPM). IPM focuses on the long-term prevention of pests through a combination of techniques—biological control, habitat manipulation, and environment modifications—with pesticides used only as needed, according to established guidelines, and applied in a manner that minimizes risks to human health (University of California, 2014). As part of this, clean refrigerators regularly, dispose of food in pest-proof bins, and remove trash daily.

For those working at home, common indoor air pollutants that might harm us include carbon monoxide, building and paint products, carpets, cleaning supplies and household chemicals, dust mites and other pests, lead, mold and dampness, pet dander, and secondhand smoke. Be mindful of these pollutants and strive to keep your home office—and your home itself—free of them.

SECURITY

PHYSICAL SECURITY

Our own personal safety, and the security of our belongings, can be a concern if we work in offices that are open to the public, especially if we are alone and working during unusual hours. In urban areas of the United States, most office buildings have formal surveillance and doorman/security systems. Limited and supervised entry, with photo identification required for visitors, is now the norm. For small offices or home offices, planned entrance and exit routes for visitors that are easily viewed by the office occupant are helpful for maintaining security. Of course, a reliable alarm system that brings help quickly is essential.

Insider theft—employees stealing from one another—is a separate problem altogether and one that should be covered via internal policies and procedures. Having safe places to lock personal and company property can help.

WI-FI SECURITY

Wireless Connections at Work

For the corporate office, we will assume here that the company has an information technology (IT) department, or independent IT provider, that sets up wireless networks and ensures their security. Briefly put, as with the home office wireless (described in the following) Internet security is commonly related to the Internet provider the company selects and protections put on the Internet to keep it safe from hackers. This will evolve as Wi-Fi technology continues to progress.

Corporate wireless networks are generally subject to security breaches in several ways. An unauthorized person can simply walk into the building and plug his or her laptop into one of the Ethernet ports connected to the company router. If a company's wireless service is unsecured, a would-be hacker can search for unsecured networks or, if it is secured, could obtain passwords from an employee or tease out passwords through devices like phishing or Trojan horses. By the time this book is published, any manner of additional hacks may exist, with what we hope are countervailing security measures.

Home Office Wi-Fi

With today's technology, setting up a Wi-Fi system in your home office is virtually a basic need. Consider the demands you will place on your Wi-Fi system—if it goes down in the middle of the workday, so does everything else, assuming your work requires constant or near-constant connectivity, which many of our work situations do. At this point in time, two related issues affect Wi-Fi reliability: the company connecting the signal and the amount of work you want the wireless system to do in your home.

When you choose the company to connect the signal, remember that when the company experiences an outage, you will as well, so select a provider with a strong track record. Securing your Wi-Fi connections is an important element of setting up your home office and securing your personal data. A Wi-Fi network using Wi-Fi protected access technology provides both security (you can control who connects) and privacy (nobody else can read or access the transmissions) for any communications across your network.

Most Wi-Fi equipment is shipped with security disabled to make it very easy to set up your network. You are given a default network name (SSID) and administrative credentials (username and password) to make configuration as simple, and once you have set up your network, you should immediately change those default settings. Consider adding other measures to secure your communications after they travel beyond your Wi-Fi network, including personal firewalls, virtual private networks, and https. These are all tools that reduce the risk of compromised privacy and security for your Internet connections. Be aware of hacking attempts through malware, Trojan horses, unsolicited ".exe" files, phishing, and other devious-minded efforts. These types of attempts will continue to evolve as technology progresses, so be sure to monitor what the latest security measures are.

Remote Wi-Fi

If you are using a handheld or laptop in a remote or on-the-go location, you should make sure your Wi-Fi devices are secure as well. Here are some general guidelines; again, these will change as technology progresses:

- *Enable whatever "wireless protected access technology" is the most current for security* (currently WPA2): All of your Wi-Fi client devices (laptops, handsets, and other Wi-Fi-enabled products) should use this.
- *Configure to approve new connections*: Many devices are set by default to detect and automatically connect to any available wireless signal. Configuring your device to request approval before connecting gives you greater control over your connections.
- *Disable sharing*: Your Wi-Fi-enabled devices may automatically enable themselves to start sharing/connecting with other devices when attaching to a wireless network. File and printer sharing may be common in business and home networks, but you should avoid this in a public network such as a hotel, restaurant, or airport hotspot.

Do you travel for business? Ever wonder which popular tourist definitions are the most risky to your cybersecurity? According to Skycure, a company that specializes in mobile threat defense, the riskiest places to use Wi-Fi among top tourist destinations are Times Square in New York City, followed by the Las Vegas Strip, Chicago's Navy Pier, and Notre Dame Cathedral in Paris.

CABLING AND WIRING

The need for data transmission, telephone lines, and electricity outlets per workstation has increased tremendously with the magnified utilization of electronic communication. It is now common to see two displays, a printer that doubles as a scanner, plus the usual telephone at a regular office workstation.

Cables, cords, and wires can access the workstation through a hollow "dropped" ceiling and power poles, via hollow accessible floors, by preset ducts in floor and wall, or (with the danger of tripping) as flat bundles on the floor or under the carpet. At the workstation, cables and wires can run inside stationary furniture or along its side—but at some point, they surface to connect to the equipment.

While providing cables, cords, and wires seems to be mainly a task for management, architect, and builder, it affects the employee in terms of safety by intruding into the workspace (no stumbling, please), it can reduce orderliness and cleanliness of the workplace, and it can adversely affect the ability to perform tasks. It impacts the general appearance of our workplace and consequently our attitude.

Of note is that increasing numbers of the devices we use are now wireless or at least able to operate on a quicker charge. We imagine that eventually cables and wiring will vanish from our offices altogether. For now, though, let us keep those cables and wires tucked away where we do not trip over them in our rush to complete a task.

FILING AND KEEPING PAPER RECORDS

A paperless office has long been promised. The 1980s gave us a first false start when electronics proliferated and progressed and the need for paper files should have abated. In actuality, with the ease of printing, paper records seemingly became even more voluminous during that time than a few decades before, which led to a need for large storage spaces for paper files—often messy and always flammable. Today's computers have far more storage capacity than ever before, and with the debut of the cloud, virtual storage capacity is almost infinite. Additionally, thumb-sized flash drives can carry massive numbers of files while fitting easily into a tiny pocket for unrestricted portability. Apps are augmenting software and are, by nature, portable, and file sharing via electronics is simple. It looks like the paperless office really may have arrived.

There may be one small drawback to the eventual demise of the old office hulk—the filing cabinet. Without files stored away from where they are needed, we do not need to periodically get up and move around to retrieve a file. This gives us all the more reason for us all to force ourselves to leave our seats and take breaks...!

COMMUNICATING

The ways in which we communicate with others have changed and will continue to change. In just a few decades, the telephone made the telegraph obsolete, pneumatic mail delivery within a building has vanished, telegrams and the telegram style verbiage are all but forgotten, and the amount of "snail mail" delivered by the postal services to addresses all over the globe was sharply curtailed by fax and e-mail, then by text, post, tweet, link, and pin. New ways to communicate change the tasks and talents of office personnel: consider, for example, that there are hardly any secretaries left who take dictation via shorthand; instead, most of us do our own word processing and information documentation and transmission via keyboard, voice, or other input device. Further development in electronic information storage and transmission will demand that we develop new facilities and skills.

FIGURE 7.1   The old-fashioned telephone—a pain in the shoulder and neck. (Courtesy of Herman Miller, Inc., Zeeland, MI.)

The telephone is one old office tool that is still in wide use. No mistake about it, it has morphed—becoming all one piece, acquiring push buttons, dropping its cord, shrinking, growing, going wireless—and its quality of transmission has improved tremendously. Yet one design flaw perseveres: it is still difficult to hold against our ear and close to our mouth simultaneously, as Figure 7.1 illustrates. This is particularly true when we talk while keyboarding or using our hands for another manual task—and, in the process, we cradle our phone between shoulder and ear, craning our neck and twisting our shoulder to keep the receiver snugly in place. This is where a headset can be useful—situating this device on our head, probably with a separate mike close to the mouth and the speaker at one ear, avoids our twisting the upper body, frees both hands, and makes for clearer communication.

Communications are covered in much more depth in Chapter 2.

OFFICE AESTHETICS

Aesthetics are important. A well-laid-out office is appealing, and may actually improve productivity and job satisfaction—Chapter 4 provides greater detail. Everyone prefers a well-furnished, appealingly colorful, well-lit, clean, and orderly office to an ugly, dark, dirty, and cluttered one. Most of us like to give our corporate workspace as much of a personal touch as possible by rearranging fixtures and furniture to our preferences, adding colors and plants, decorating with pictures and photos, and bringing in items we cherish. This helps us feel well at work and feel good about work, and—directly or indirectly—to work with ease and efficiently (yes, ergonomically).

The same is true for your home office—most of us are more comfortable and more efficient if we keep our workspace clear and free of distractions.

Josh recently left his job as a corporate financial analyst to work as a private tax preparer and personal financial consultant. To launch his business, he reached out to former colleagues and friends in person and via social media and between that and word of mouth, he has already attracted 15 clients. Josh also has a 13-year-old son and two active dogs. He has set up his work at the kitchen table—no point in investing in a home office until he has more established! This morning, after dropping his son off at school, he settles in to review his upcoming workload. The kitchen table is next to a window, and Josh gazes outside as he opens his laptop and shuffles through some client paper folders next to the computer. His eyes wander over piles of snow in the yard and tree boughs bent with ice. It is winter, and tax season is rapidly approaching. It seems like everyone of his clients wants tax preparation assistance, which is great for his business. But Josh knows it also means he needs to focus and get to work. He cannot help but notice that his son's breakfast dishes are still on the table in front of the computer—plate sticky with maple syrup and pancake remnants, half empty glass of milk leaving damp rims on the table, and fork dripping yellow egg from its tines. Should he clean it up before the egg on the fork congeals and before the glass leaves permanent damage to the table top? And he could have sworn that his son did not remember to grab the school lunch in the fridge. Should he run to school and drop it off? Josh sighs and pushes his chair away from the table. The dogs could use a walk. And his son far prefers homemade lunches to what they dish out at the cafeteria. He will just tidy up the breakfast mess and walk the dogs over to his son's school to drop off the lunch. But, he reminds himself, he will have to truly concentrate on work when he returns.

SEDENTARY OFFICE WORK

We have mentioned the dangers of sedentary work in several of the chapters of this book, and we raise the topic again here, because too much sitting is, quite simply, bad for us, and consequently an office hazard. Office workers spend a large part of their day sitting down, working at their keyboards, and this adversely affects their cardiometabolic and musculoskeletal health.

A new trend in office ergonomics involves replacing traditional desks with dynamic workstations, ones where work can be done while standing, walking, or biking. These include sit-to-stand tables, treadmills, stationary bicycles, and exercise balls. Changing our posture often is good for us, and the most effective workstation appears to be the sit-to-stand table when it comes to decreasing musculoskeletal pain.

Of course, changing from sitting to standing does not completely eliminate health issues, because prolonged standing can also be painful over time. However, it is likely that any workstation that reduces the amount of sitting and encourages rotating between standing, sitting, and possibly other activities will result in better health outcomes and improved comfort. And it is not just the actual workstation that should encourage movement, but the overall design of the office should do so as well (see Chapter 4).

Importantly, having an "active" workplace also needs to be encouraged by management and supported through training and communications. Replacing traditional workstations with dynamic ones will not do much if employees are not using them; the company needs to educate employees on why movement is important, teach them how to use the workstations, and offer ongoing training and support to encourage these healthful behaviors. Ideally, dynamic workstations would be one component in a whole system of employee health and wellness programs at work.

SHIFT WORK

Shift work—defined here as any work shift that veers significantly from the traditional "9 to 5" —is a reality for many employees. Over the past few decades, many countries have become increasingly dependent upon shift workers to meet the demands of globalization and our 24-hour society. From a competitive standpoint, shift work is an excellent way to increase production and customer service without major increases in infrastructure. Millions of Americans are considered shift workers, including physicians and nurses, pilots, bridge builders and road construction crews, police officers, shipping service workers, call center employees, tech and information technology representatives, customer service representatives, and commercial drivers (CDC, 2015).

Shift work has both advantages and disadvantages, and it fits some individuals better than others. In general, working the second (evening) or third (night) shift means more pay than working the day shift; and some fans of shift work appreciate that better jobs are sometimes available to them at night because there is not as much competition for these positions. This can also mean that second and third shift workers get promoted more quickly than their daytime colleagues might. Some third shift workers also describe a looser, friendlier work environment at night, with more camaraderie and a shared focus on completing the work. Night shifts are also usually less busy than a regular day shift, with less traffic and distractions. There are other intangibles, like greater autonomy and fewer meetings (management is sleeping).

How a person copes with shift work, however, varies by the individual. In 2008, the U.S. National Library of Medicine conducted a study that showed that, with regard to coping and adjusting, "adaptive" attitude was more important than the hours worked (McCarten, 2011). Also, other factors come into play when it comes to the suitability of shift work. Consider parents with small children. One parent may opt for permanent evening or night shift work so that, if the other parent works days, someone can always be home with the kids for 24-hour childcare. This reason—childcare—is a compelling one: according to a study from California more than a fourth of night shift workers choose that schedule because of childcare needs (Foster, 2014).

The drawbacks of shift work are obvious—shift work inherently disrupts natural sleep–wake patterns and this can lead to a host of troubles. Let us take a closer look at the sleep–wake cycle: it appears to have evolved so that humans are awake during the day and asleep for approximately 8 hours at night. The brain contains a functional set-up called the "circadian clock," which monitors the amount of light we see in each moment. When the light begins to fade, in the evening, the clock allows the release of melatonin, a brain chemical that gives the body the signal that it is time for sleep. Overnight, melatonin levels remain high. At daybreak, they drop, and during the day, they stay low.

The circadian clock is active in other ways, too. It dispatches other brain chemicals, like noradrenaline and acetylcholine, to keep us awake during the day, and it plays a role in body temperature, heart rate, blood pressure, and even digestion. All of this activity over the course of a 24-hour day is called the circadian rhythm, and it functions without any conscious or deliberate intervention.

Now, imagine you have taken on a new job working the third shift. This means you will begin at 9 in the evening and you will complete your "day" at 6 in the morning. To function and perform, you will need to override what your body is naturally programmed to do. Simply put, your routine will run counter to your circadian rhythm, and your body might not like that—and develop health problems as a result.

Research findings are beginning to show that shift work can be hazardous to your health. Specifically, shift work has been linked to

- Sleep disorders
- Increased likelihood of obesity
- Increased risk of cardiovascular disease
- Higher risk of mood changes and disorders
- Increased risk of digestive and gastrointestinal problems
- Increased risk of other diseases
- Higher incidence of motor-vehicle and work-related accidents and errors

We will discuss each possible consequence here.

SLEEP DISORDERS

It is no surprise that many shift workers suffer from sleep disorders. Shift workers get, on average, 2–3 hours less sleep than other workers (UCLA Health, 2015). When they do sleep, during the day, it is often split into two different sections, with a few hours in the morning, and then a shorter nap in the afternoon before going to work. Because of our circadian rhythms and the natural daylight during the day, it is quite simply more difficult to sleep.

Some shift workers suffer from a more serious daytime sleepiness, called shift work sleep disorder (SWSD). SWSD consists of recurring or constant sleep issues in shift workers, marked either by difficulty sleeping or by excessive sleepiness. Other symptoms include lack of concentration, headaches, and drained energy. Consequences of this disorder include increased accidents and work-related errors, increased sick leave, and irritability or mood problems. Indeed, patients with sleep disorders spend twice as many days in bed per year due to sickness and miss 3–4 more days at work (Bajraktarov et al., 2011).

If symptoms persist, individuals should seek medical advice, but there are some general guidelines those suffering from SWSD can follow to help mitigate their symptoms. These include minimizing exposure to light on the commute home to keep sunlight from activating the circadian clock, following bedtime rituals and maintaining a regular sleep schedule, going to sleep as soon as possible after work, and keeping a quiet, dark, and peaceful environment during sleep times. Blackout curtains can help, family members should keep any noise to a minimum, and your cell phone should be off. Experts also advise that shift workers keep up the sleep schedule on days off, and if all else fails, prescription medication to promote wakefulness and sleep aids to fall asleep can be used (WebMD, 2014).

INCREASED LIKELIHOOD OF OBESITY, DIGESTIVE ISSUES, AND CARDIOVASCULAR DISEASE

A link to obesity, digestive issues, and cardiovascular disease can develop with insufficient sleep and disruptions to the circadian rhythm. The circadian rhythm guides bodily functions such as heart rate, blood pressure, body temperature, digestion, and brain activity, so ignoring—even defying—this natural rhythm throws those functions off. Insufficient sleep has long been known to adversely affect metabolism and appetite, and studies have shown that shift workers have higher levels of triglycerides than day workers. They have also shown that shift workers suffer significantly more upset stomachs, ulcers, and bouts of constipation and indigestion than day workers do (HealthDay, 2015). In addition to these biological factors, we must add in lifestyle realities—shift workers often have irregular eating habits and easy access to poor food choices (fast food, say) along with fewer healthy options (farmers' markets, for instance). All of these increase the risk of metabolic problems.

HIGHER RISK OF MENTAL HEALTH ISSUES

Shift work changes our routine and disrupts our social interactions which can cause us to feel unhappy. Let us go back to imagining that you have accepted a job on the third shift and your workday begins late in the evening. Your family and friends go about their regular routine: they are still planning family dinners, and 5K fun runs on Saturday mornings, and cocktail hour on Wednesdays to celebrate #WineWednesday, but you cannot go, because your main meal is during the early afternoon, you are just starting your sleep time during those 5K mornings, and if you had a cocktail to celebrate #WineWednesday, you would be starting your shift with alcohol in your system. Frankly, we would feel glum about that, too.

Shift work disorder can increase the risk of mental health problems, like mood disorders and depression. Again, disrupting the circadian system means disrupting the ebb and flow of brain hormones, and this affects our mood. Alon Avidan, Director of the UCLA Sleep Center, opines that shift workers accumulate sleep debt and, as a result, are at greater risk of developing psychiatric conditions. "Things get to a point where it begins to impact their social function and relationships." he says (Foster, 2014).

Increased Risk of Other Diseases

Long-term physical health effects associated with working nights are well documented, but there are some links that are still not completely understood. Experts believe that suppressed levels of melatonin, a hormone that is usually produced at night, could be to blame. Melatonin regulates pituitary and ovarian hormones—including estrogen—and elevated levels of estrogen are linked to increased risk of reproductive cancers.

Increased Incidences of Errors and Accidents

Errors at work and even accident probability can also increase for shift workers. As mentioned earlier, shift workers get an average of 2–3 hours less sleep every night than nonshift workers. Since they are less likely to sleep the full amount their bodies' require, they can amass a large "sleep debt" over time, and that can mean slowed reaction times, delayed thinking, and lowered ability to solve problems. It might also be harder to stay focused and on-task. These issues are exacerbated by the disruptions to circadian rhythms, as described earlier. For those individuals working in professions where mistakes come with potentially dire consequences—physicians and nurses, pilots, truckers, and heavy machinery operators—the potential for danger is very real. Kroemer (2016) summarizes research that shows increased safety risks for shift workers: those who work nights are more likely to report nodding off at work and "drowsy" driving to and from work. Consider that major accidents in recent history—the Exxon Valdez oil spill, Three Mile Island, and Chernobyl meltdown—all took place during overnight shifts.

OCCUPATIONAL DISEASES AND KEYBOARDING

Keyboarding—whether done on a typewriter, calculator, desktop, laptop, or notebook computer and whether the keyboard itself is manual, virtual, wireless, oversized, and undersized—can lead to various kinds of bodily discomfort: often in the hands and wrists or in the shoulders. Chapter 6 covers keyboarding and the issues related to keyboarding extensively. Overuse disorders related to keyboarding can be damaging and long term-see Appendix C for more information.

An overuse disorder (OD) can present itself slowly over days or weeks or months, or it may appear fairly suddenly, even if you enjoy the activity that you are overdoing, such as knitting, playing the piano, or even swinging your golf club. First signs often appear as slight pains and occasional aches; if the issues are in your hand, you may experience numbness and tingling. Continuous pain then may ensue, often with symptoms of inflammation and swelling, especially of tendons and their sheaths. Overuse disorders, include sets of long-known health problems-see below for other names and details. They can be averted by using more appropriate and better-fitting equipment and, most importantly, by not "overworking" the body. When you do experience the initial signs of an overuse injury, stop doing what is causing the problem, talk with your supervisor, and get medical treatment as needed.

OCCUPATIONAL DISEASES

Three hundred years ago, Bernardino Ramazzini, "the father of industrial hygiene," wrote about occupational injuries in offices (if you are interested, please see Wright's 1993 translation of the Latin text). Of course, in the early 1700s, there were no typewriter or computer keyboards yet, but muscle pain and cramps were prevalent among secretaries and scribes, stemming from repetitive tasks that involved long-maintained hand and arm postures. Ramazzini also described many other occupational diseases, and in the late-1800s, overuse disorders were well known to be correlated with certain occupations, for example, among textile workers or among musicians, especially pianists and then among "keyboarders" working on typewriters and computers (Kroemer, 2001). More information on early discussions of ODs can be found in Appendix C.

In the 1990s, a large variety of occupation-related ODs were described in the literature, compiled in much detail by Kroemer in 2001. The factors that cause, aggravate, or precipitate ODs can be part of occupational or leisure activities, as indicated by such evocative and descriptive names as

- Writer's cramp or scribe's palsy
- Goalkeeper's or seamstress' or tailor's finger
- Bowler's or gamekeeper's or jeweler's thumb
- Bricklayer's hand
- Meatcutter's or musician's or pianist's or stitcher's or tobacco primer's or telegraphist's or washerwoman's wrist
- Carpenter's arm
- Carpenter's or jailer's or student's elbow
- Porter's neck
- Shoveler's hip
- Weaver's bottom
- Housemaid's or nun's knee
- Ballet dancer's or nurse's foot

In the twentieth century, this list was extended with new descriptive terms: baseball catcher's hand; typist's, cashier's, letter sorter's, and yoga wrist; golfer's or tennis or mouse elbow; letter carrier's shoulder; carpet layer's knee and, most recently, texter's thumb.

CTDs

The prominent reasons for such disorders are highly repetitive activities, often with awkward positions and movements of the body segments involved, and pressure from equipment on the body—such as is often associated with keyboarding. A single such small trauma is not injurious to the body if it occurs occasionally, but the cumulative effects of "microtrauma" can lead to overexertions.

Here is a formal definition: "Cumulative trauma disorders (CTDs) are regional impairments of muscles, tendons, tendon sheaths, ligaments, nerves, and joints associated with activity-related repetitive mechanical trauma." Other names used are repetitive trauma injury (RSI) (or illness, or disorder), repetitive motion injury (RMI) (or illness, or disorder), repetitive strain injury (RSI), (occupational) overuse disorder (OD), and work-related musculoskeletal disorder (WRMD).

Medically, CTDs are usually classified as irritation, sprain, strain, or inflammation. The terms used to describe their nature include tendinitis (tendonitis), tenosynovitis, myalgia, bursitis, peritendinitis, epicondilitis, brachial plexus syndrome, thoracic outlet syndrome, carpal tunnel syndrome (CTS), regional pain syndrome, neurovascular syndrome, cervicobrachial syndrome, compression syndrome, entrapment syndrome, and neuropathy.

What Was Known about CTDs and When

As Ayoub and Wittels (1989), Burnette and Ayoub (1989), Burry and Stoke (1985), Hochberg et al. (1983), Fry (1986), and Lockwood (1989) described in detail and as listed in Appendix C, activity-related causes of CTDs were extensively reported in the literature of the 19th century, and their physiological–pathological mechanisms were extensively discussed even by the end of that century.

Further insights in the pathology of CTDs were gained in the first four decades of the twentieth century. CTDs were now firmly linked to physical strain of the musculoskeletal system, with the stress generated not so much by the amount of energy expended in single muscular activations but rather by the accumulation of efforts in highly repetitive work, as explained by Schroetter (1925), Klockenberg (1926), Conn (1931) and especially by Hammer (1934). CTDs were known to be related to design features and operation of equipment, such as hand tools, telegraphs, and especially the typewriter.

Typing in the early twentieth century was rigorous work. A typist had to practically pounce on the keyboard, fingers pushing hard on resistant keys and stretching out to reach distant ones. The arrangement of the keyboard itself, and of the keys on it, forced the arms into strong inward twist (pronation) and the hands into lateral bend (ulnar deviation) and required complex motions between poorly located keys. As Appendix C shows, design deficiencies in Shole's original layout quickly became clear, and improvements were proposed as early as 1915; Heidner split the keyboard, tilted the halves down sideways, and rearranged the key layout. In 1920, Banaji, Nelson, and Wolcott all acquired patents for their rearrangements of the keys to alleviate the complexity of finger movements, with Dvorak's (1936) patent the best known nowadays.

By 1960, physiological/biomechanical knowledge causally linked repetitive activities, particularly typing, to strains in the tendons and sheaths of the hand, wrist, and arm. Lundervold (1951, 1958) demonstrated that electromyography was a useful measure of muscle efforts. Pfeffer et al. (1988) stated that, by 1960, CTS was the most frequently diagnosed, best understood, and most easily treated entrapment neuropathy.

By 1980, CTDs were generally understood to result from a series of microtraumas. These, as mentioned, would be harmless if they occurred only occasionally and infrequently. Over time, however, and with cumulative effects, they were shown to injure musculoskeletal, vascular, and nervous systems of the human body. Given this thorough understanding of human CTD pathology, avoiding repetitive motions in awkward hand/arm postures, such as in typing, had become a goal of industrial hygiene. Armed with medical knowledge, practical experiences, and specific research, ergonomists made many well-considered recommendations for new designs of keys and keyboards and for their use, as Table C.4 shows. Replacing mechanical levers between key top and platen on the typewriter with electromechanical devices began around 1970—and with electronic switches and circuitry—facilitated new designs of keys and keysets. This effectively reduced key travel and key resistance as compared to the old typewriter, but not keying frequency. However, the number of keys on all-purpose electronic keyboards in the 1980s was typically just more than 100, more than double as many as on Sholes' (1878) design.

During the 1980s, "human-engineered" keyboards and mouse input devices began proliferating, as shown in Table C.5. Several designs broke the keyboards into one section each for the left and right hand and set the keys into hand-configured arrays, often similar to those designs offered in proposals made early in the century. Using wrist rests, adjusting location and angulation of the keyboard, adopting a more ergonomic layout of the computer workstation, and taking breaks from keying were becoming popular as techniques to avoid CTDs.

During the 1990s, the biomechanical strains underlying CTDs, especially CTS, were researched in detail. Truly, there is no mystery about what causes them, how damaging they can be, and how to avoid them. Their correlation to workload on the job, especially associated with keyboarding, is well understood. However, there is still substantial resistance to change when it comes to keyboards. Proponents of keeping the traditional QWERTY layout argued that learning curves associated with new keyboard layouts would be prohibitive and output would suffer. As listed in Table C.6, the specific traits of new keyboard designs to avoid overloading of their operators were tested. It turns out that nontraditional (slit, slanted, and tilted) keyboards did not require long retraining periods and did not reduce keying performance as compared to flat conventional keyboards; in fact, use of the ergonomic keyboards resulted in better performance.

Summary of Keyboarding and Occupational Diseases

It took nearly a full century to make the general public aware that the highly repetitive action of tapping down on keys can overexert the musculoskeletal system of hands and arms. Yet research done as early as in the 1920 had already pinpointed that problem, and inventors had obtained patents for better keyboards. The energy required for each key stroke decreased dramatically when electronic computers replaced mechanical typewriters, but the basic QWERTY arrangement of the keys on the keyboard itself did not change, in spite of all the well-documented shortcomings. Instead, more keys of the same kind were added, increasing the distances that fingertips must travel.

Keying-related disorders, well known to befall many typists during the first half of the century, became rampant again in the 1980s. The burden continues to fall on the keyboard user; however, in 1996, the U.S. Court of Appeals for the Third District decided in a class-action suit that keyboard manufacturers did not have to provide warnings about possible health risks associated with the use of their conventional keyboards.

Many kinds of improved keyboards are presently on the market, but most of us still adhere to the QWERTY model. Advances in technology, paired with the recognition of the human hand's biomechanical limitations, should lead to better means of transferring information from the human to the computer.

ERGONOMIC RECOMMENDATIONS

Whether you are in a home office, a company office, or some other virtual work site, chances are the space is probably safe vis-à-vis fire hazards and indoor air quality. However, there is always a small amount of risk, and it just makes sense to minimize any dangers that may lurk unnoticed. To this end:

- Make sure you know where the exit doors are, that your fire alarm is working and (as appropriate) connected to a power or battery source, where your fire extinguisher is, and what your fire evacuation plan is (the fastest way out). If you are working in a remote location such as a café or library, look around, know where the second exit is, and familiarize yourself with the surroundings.
- Cabling and wiring should be out of your way. This includes the cables and wires under your work surface so they do not trip you if you need to get up quickly.
- Be aware of the main contributors to poor air quality and minimize your exposure to them as much as possible. These contributors include inadequate ventilation systems, water damage and mold growth, office overcrowding, cubicle design that blocks off air flow, too much or too little humidity, cleaning chemicals and pesticides that are stored incorrectly, and poor housekeeping with dust and dirt remaining in our work environment. For individuals with respiratory issues like asthma, these air pollutants are especially troublesome. To improve indoor air quality, consider the following:
  - Keeping your environment completely smoke-free.
  - Establishing or following a fragrance-free policy: perfumes, colognes, aftershave, scented lotions, or hairsprays are all disallowed.
  - Maintaining the office or home HVAC system and ensuring it is working well.
  - Taking flu shots: prevention is the best medicine when it comes to influenza.
  - Keeping pests at bay.

From a security perspective, both your physical security and your wireless security are relevant.

- Many urban office buildings have formal surveillance and doorman/security systems; if you are in a building that does not have these features, be particularly vigilant.
- For wireless security, your company should ensure that the service is secured; if you are using personal Wi-Fi, or sharing Wi-Fi in a public setting (i.e., a café), enable whatever "wireless protected access technology" is the most current for security purposes, and reconsider sending or accessing sensitive information.

Take frequent breaks and get up and move around, even if you are working from a remote location. If your workload does not allow for frequent breaks, consider a "dynamic" workstation—one that allows you to shift frequently between standing and sitting or permits other kinds of movement, like walking or biking.

If you use the telephone often or over long periods, consider a headset. If you do repetitive work such as keyboarding, do it correctly and take frequent breaks. Experiment with ergonomic keyboards instead of a flat one-piece conventional keyboard; there is a learning curve associated with adapting to a new keyboard, especially if it is very different than the traditional one, but we tend to adapt surprisingly quickly.

Make your workplace appealing. Rearrange furniture and devices; bring in color and light to your taste. Most of us are easily distracted, so minimize distractions as much as possible.

If you are one of the many people around the globe who do shift work—any work shift that veers significantly from the traditional "9 to 5"—be aware of the risks associated with nontraditional work schedules. Shift work counteracts your natural circadian rhythm and inherently disrupts natural sleep–wake patterns; this can lead to a host of troubles. Specifically, shift work has been liked to

- Sleep disorders
- Increased likelihood of obesity
- Increased risk of cardiovascular disease
- Higher risk of mood changes and disorders
- Increased risk of digestive and gastrointestinal problems
- Increased risk of other diseases
- Higher incidence of motor-vehicle and work-related accidents and errors.

If you find yourself struggling with any one or more of these symptoms, consider making a change in your schedule.

Do not hesitate to make well-thought suggestions for improvements in work procedures, equipment, tools, and appearances.

REFERENCES

ACAAI. (2014). Who has asthma and why. American College of Allergy, Asthma, and Immunology. Arlington Heights, IL: http://acaai.org/asthma/who-has-asthma. Accessed November 2015.

American Academy of Sleep Medicine. (2014). Shift work—Overview. *Sleep Education*. http://www.sleepeducation.org/essentials-in-sleep/shift-work. Accessed November 2015.

American Lung Association. (2015). Indoor air quality. [Blog]. Toxic free consumer. http://toxicfreeconsumer.com/wp-content/uploads/2015/09/Indoor-Air-Quality-American-Lung-Association.pdf. Accessed November 2015.

Ayoub, M. A. and Wittels, N. E. (1989). Cumulative trauma disorders. *International Reviews of Ergonomics* 2, 217–272.

Bajraktarov, S., Novotni, A., Nensi, M., Dance, G., Miceva-Velickovska, E., Zdraveska, N., Samardjiska, V., and Richter, K. (2011). Main effects of sleep disorders related to shift work—Opportunities for preventive programs. *The EPMA Journal* 2(4), 365–370.

Banaji, F. M. M. (1920). Keyboard for typewriters. Patent 1,336,122. United States Patent Office, Alexandria, VA.

Burnette, J. T. and Ayoub, M. A. (July–August 1989). Cumulative trauma disorders. Part 1. The problem. *Pain Management* 2, 196–209.

Burry, H. C. and Stoke, J. C. J. (1985). Repetitive strain injury. *New Zealand Medical Journal* 98, 601–602.

CDC. (2015). Work schedules: Shift work and long hours. National Institute for Occupational Safety and Health. http://www.cdc.gov/niosh/topics/workschedules/. Accessed November 2015.

Conn, H. R. (September 1931). Tenosynovitis. *The Ohio State Medical Journal* 27, 713–716.

Drake, C. (2015). Shift work and sleep. National Sleep Foundation. Washington, DC: https://sleepfoundation.org/sleep-topics/shift-work-and-sleep. Accessed November 2015.

Dvorak, A. (1936). Typewriter keyboard. Patent 2,040,248. United States Patent Office, Alexandria, VA.

Foster, B. (2014). The night shift. *Psychology Today.* https://www.psychologytoday.com/articles/201404/the-night-shift. Accessed November 2015.

Fry, H. J. H. (1986a). Overuse syndrome in musicians—100 years ago. A historical review. *The Medical Journal of Australia* 145, 620–625.

Fry, H. J. H. (1986b). Physical signs in the hands and wrists seen in the overuse injury syndrome of the upper limb. *Australian and New Zealand Journal of Surgery* 56, 47–49.

Fry, H. J. H. (September 27, 1986c). Overuse syndrome in musicians: Prevention and management. *The Lancet* 2, 723–731.

Fry, H. J. H. (1986d). What's in a name? The musician's anthology of misuse. In *Medical Problems of Performing Artists and Incidence of Overuse Syndrome in the Symphony Orchestra.* Philadelphia, PA: Hanley & Belfus, pp. 36–38, 51–55.

Hammer, A. W. (October 3, 1934). Tenosynovitis. *Medical Record* 140, 353–355.

HealthDay. (2015). Shift workers. *Occupational Health News.* http://consumer.healthday.com/encyclopedia/work-and-health-41/occupational-health-news-507/shift-workers-646677.html. Accessed November 2015.

Heidner, F. (1915). Type-writing machine. Letter's Patent 1,138,474. United States Patent Office, Alexandria, VA.

Hochberg, F. H., Leffert, R. D., Heller, M. D., and Merriman, L. (1983). Hand difficulties among musicians. *The Journal of the American Medical Association* 249(14), 1869–1872.

Hoffman, J. (2014). Sleepless in America [Documentary]. National Geographic Channel. National Geographic. Public Good Projects, National Institutes of Health. Washington, DC.

Klockenberg, E. A. (1926). *Rationalization of the Typewriter and Its Operation* (in German). Berlin, Germany: Springer.

Kroemer, K. H. E. (2001). Keyboards and keyboarding. An annotated bibliography of the literature from 1878 to 1999. *International Journal Universal Access in the Information Society* 1(2), 99–160.

Lockwood, A. H. (1989). Medical problems of musicians. *The New England Journal of Medicine* 320(4), 221–227.

Lundervold, A. (1958). Electromyographic investigations during typewriting. *Ergonomics* 1, 226–233.

Lundervold, A. J. S. (1951). Electromyographic investigations of position and manner of working in typewriting. *Acta Physiologica Scandinavica* 24(Suppl. 84), 1–171.

McCarten, K. (2011). Coping with the challenges of shift work. *Nursing News and Events.* http://www.nursezone.com/Nursing-News-Events/more-news/Coping-with-the-Challenges-of-Shift-Work_37124.aspx. Accessed November 2015.

Murray, P. (2014). Sleepless in America review. *Sleep Education.* American Academy of Sleep Medication. http://www.sleepeducation.org/news/2014/12/02/sleepless-in-america-review. Accessed November 2015.

Nelson, W. W. (1920). The improvements in connection with keyboards for typewriters. British Patent 155,446.

Occupational Safety and Health. (2007). Fire safety. OSHA. United States Department of Labor. Washington, DC: https://www.osha.gov/SLTC/firesafety.

Pfeffer, G. B., Gelberman, R. H., Boyes, J. H., and Rydevik, B. (1988). The history of carpal tunnel syndrome. *Journal of Hand Surgery* 13-B, 28–34.

Schroetter, H. (1925). Knowledge of the energy consumption of typewriting (in German). *Pflueger's Archiv fuer die Gesamte Physiologie des Menschen und der Tiere* 207(4), 323–342.

Sholes, C. L. (1878). Improvement in type-writing machines. Letter's Patent 207,559. United States Patent Office, Alexandria, VA.

UCLA Health. (2015). Coping with shift work. UCLA Health Sleep Disorders Center. http://sleepcenter.ucla.edu/body.cfm?id=54. Accessed November 2015.

University of California. (2014). Definition of integrated pest management. http://www.ipm.ucdavis.edu/GENERAL/ipmdefinition.html. Accessed November 2015.

WebMD. (2014). Sleep and circadian rhythm disorders. http://www.webmd.com/sleep-disorders/guide/default.htm. Accessed November 2015.

Wi-Fi Alliance. (2013). Security. http://www.wi-fi.org/discover-wi-fi/security#sthash.y64O5guJ.dpuf. Accessed November 2015.

Wolcott, C. (1920). Keyboards. Letter's Patent 1,342,244. United States Patent Office, Alexandria, VA.

Wright, W. C. (1993). *Diseases of Workers, Translation of B. Ramazzini's 1713 "De Morbis Articum"*. Thunder Bay, Ontario, Canada: OH&S Press.

8 Seeing and Lighting

The baby bat screamed out in fright, "Turn on the dark I'm afraid of the light!"

—**Shel Silverstein, poet, screenwriter, and author**

Stars and Shadows ain't good to see by.

—**Mark Twain, *The Adventures of Huckleberry Finn***

OVERVIEW

It Is Simple: With Light, We See; Without, We Cannot

In order for us to see, there must be light. Objects must "appear bright" so that we can see them in detail; when light dims, objects lose their color, and when they are enshrouded in dark, we cannot see them at all. Even a well-lit visual target must be at the correct distance from our eyes so that we can distinguish particulars. If our eyesight is not perfect, and as we age, it is especially important to have good lighting to carefully locate things that we must see, and we may need to refine or correct our vision with artificial lenses.

Human factors engineers know very well how to set up lighting in the office so that visual work tasks are easy to do. They also know where certain kinds of visual targets (such as written material or computer displays) should be located to make our work as easy and efficient as possible. Details are described in the following text.

In spite of the availability of both human factors professionals and existing literature on the subject of lighting, people still make surprising numbers of fundamental mistakes in selecting and placing visual targets and in setting up office illumination. This chapter gives straightforward recommendations on how to do it right, and we explain the science behind eyesight and illumination to make these guidelines logical.

EYE MUSCLES HELP US TO SEE

Whenever we direct our sight to a new target, muscles turn our eyeballs (and possibly our head, neck, and even the upper trunk) in that direction. If the object we are trying to see is in the midst of rapid motion or if we are scanning something quickly, as in reading text, our eyes must move quickly as well, and this movement is achieved by muscles rotating both eyeballs in the same way. As we switch from distant to close vision, muscles rotate the eyeballs slightly against each other so that the lines of sight of our two eyes converge. Still other muscles adjust the thickness of the lenses within our eyes. Even more muscles adjust pupil sizes as the visual environment progresses through different lighting levels.

ILLUMINATION, LUMINANCE, AND VISION

Too much—or improper—light can actually reduce our ability to see. If a light source like a bright lamp shines straight into our eyes, we experience direct glare that, if strong, might make it impossible to see anything else. Daylight and lamps put visible light energy into the office. The flow of light shines onto the objects in its path; they become illuminated. Our eyes do not see the light that falls on an object (incident light is called illumination or illuminance); instead, we see the light that is reflected from the object toward our eyes. Consider two colored surfaces, side by side, in the same flow of incident light so that they are equally illuminated: the darker surface absorbs much of the illumination and reflects only a little, while the light surface reflects most energy and absorbs little. The amount of reflected energy makes objects appear dark or light.

This reflected visible energy is called luminance. What determines how well we can see a visual target and distinguish its features, such as our colleague's face, a written text, or the computer display, are the contrasting luminances of the target's details. Visibility relies on the differences in luminance of the parts of the face that are brightly lit or shaded or, put differently, on the luminance contrast between dark and light. Consider the everyday occurrence of reading written material: black letters against their pale background on either paper or the computer screen are generally easily visible to us because of the luminance contrast between dark and light at their outlines.

Glare ranges in intensity: it may be simply a source of discomfort, or it may actually make seeing well a sheer impossibility. Glare occurs if sources of light (such as a blazing sunny window or an overly bright lamp—or even their reflections) are much brighter than the environment to which the eyes are adjusted. If the sun shines into your eyes, you probably cannot read the relatively dim text on your screen; if a high-energy light is reflected by the screen into your eyes, the text on the monitor does not show enough contrast for you to read it.

FEELING COMFORTABLE

What we consider comfortable is in line with the lighting engineer's design goal: we want an office environment that

- Allows us to clearly and pleasantly see what we want to see
- Prevents glare and annoying bright spots in our visual field
- Pleases us in terms of contrast and colors

Note: You may skip the following part and go directly to the "Ergonomic Design Recommendations" section at the end of this chapter—or you can get detailed background information by reading the following text.

HOW WE SEE

While the physiology of our eyes is relatively easy to understand, the physics of vision and lighting is more complex. We cover the basics of vision and lighting in the following text.

ARCHITECTURE OF THE EYE

How we humans see, and how our eyes function, is thoroughly researched. The *eyeball* is a roughly spherical organ of about 2.5 cm in diameter, surrounded by a fibrous layer called the "sclera," which helps the eyeball retain its round shape. When a beam of light from a distant target reaches the eye, it first passes through the *cornea*, a translucent bulging round dome at the front of the eyeball, kept moist and nourished by tears. Then, the light runs through a chamber filled with the *aqueous humor*, a clear, watery fluid that flows between and nourishes the lens and the cornea. Behind it is the *iris*, the tissue surrounding a round opening called the *pupil*. Muscles open and close the pupil like the aperture diaphragm of a camera, regulating the amount of light that enters the eye.

After passing through the pupil, the light beam enters the *lens*. If the visual object is at a distance, ligaments keep the lens thin and flat so that incoming light rays are not bent. For close objects, the ciliary muscle around the lens makes it thicker and rounder so that the light beams are bent ("refracted" is the commonly used technical term) for suitable focus. In the young and healthy eye, the cornea, aqueous humor, and lens refract incoming light beams so that they come to a sharp focus on the *retina*, the innermost layer in the rear of the eyeball.

The beam of light then passes through a large space behind the lens: it is filled with the *vitreous humor*, a gel-like fluid. The light finally reaches the retina, a thin tissue that lines about three-quarters of the inner surface of the eyeball at its rear. The retina carries nearly 140 million *light sensors* that react to light energy (Figure 8.1).

There are two kinds of light sensors on the retina, named for their shape. The majority of the sensors, about 120 million of them, are *rods*, which respond even to low-intensity light and provide black–gray–white vision. The other sensors are *cones*, which respond to colored light—provided that it is intense enough. Cones are concentrated in a small area of the retina, called the *fovea*. It is located in the spot

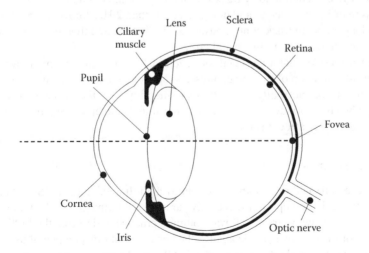

FIGURE 8.1   Sketch of the eye.

where a straight line, called the visual axis, coming through the centers of cornea, pupil, and lens, strikes the retina. Each cone contains a pigment that is most sensitive to either blue, green, or red wavelengths. Chemical reactions in cones and rods create electrical signals that are passed along the optic nerve to the brain.

MUSCLES OF THE EYE

Several groups of muscles adjust and move the eye. Two muscle groups are inside the eyeball for internal adjustments. One muscle pair is in the iris where it controls the size of the opening of the pupil. The other muscles are at the lens. The lens is normally held in a stretched position by the suspensory ligaments so that the eye is focused on far objects. When focusing on a closer object (accomodation) is necessary, the ciliary muscles overpower the ligaments and make the lens thicker. The elasticity of the lens usually declines with aging, so most people start to realize sometime around their forties that they need artificial lenses for seeing close objects. These manufactured lenses take over the job that the now stiff natural lenses can no longer do. In the United States, these glasses are commonly known as "reading glasses."

Other muscle groups are outside the eye. Six muscles attach to the outside of the eyeball controlling its movements in *pitch* (up and down), *yaw* (left and right), and *roll* (turning clock- or counterclockwise), all around a common center of rotation located approximately 13.5 mm behind the cornea. These muscles determine the mobility of the eyes. This mobility entails primarily two kinds of movements: when one follows an object that moves along a path that stays at the same distance, the lines of sight of a pair of eyes remain parallel. This is called a conjugate movement. The other is a verging movement, where the lines of sight of a pair of eyes do not remain parallel but meet at the visual target. As it comes closer, the eyes converge; as it moves away, they diverge.

The eye can continuously track a visual target that is moving left and right (yaw) at less than 30° per second and doing so at less than 2 Hz. Above these rates, the eye is no longer able to track continuously but lags behind and then must move in jumps (saccades) to catch up to the visual target.

Other muscles that are crucial for vision include the muscles that move the head on the atlas (the top of the spine), muscles that bend the neck, and even muscles that twist the trunk. Normally, they all work smoothly with the muscles of the eyes to bring objects into view, but injury or aging might influence their functioning and, hence, adversely affect vision.

LINE OF SIGHT

If the eye is fixated on a point target, the straight line of sight (LOS) runs from the object through the pupil to the receptive area on the retina, most likely the fovea.

Until just a few years ago, it was common practice to describe the LOS direction in front of the eye by its angle against the horizon. But this practice of using a reference external to the body obscures the fact that we unconsciously adjust not only the location and direction of the eyeballs but, at the same time, the postures of the head,

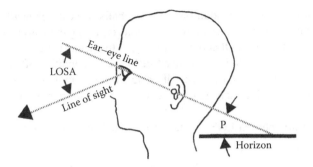

FIGURE 8.2 The ear–eye (EE) line, its pitch angle P against the horizon, and the angle of the line of sight against the EE line.

the neck, and often even the upper trunk (Delleman, 2000). What we must know is the line-of-sight angle (LOSA) against the head, not against the horizon, in order to place visual targets in the workstation for both efficient and comfortable viewing. To determine the appropriate LOS and its LOSA, the ergonomist needs a reference that moves with the head, not one that is independent of head motion.

An easy-to-establish reference on the head is the ear–eye (EE) line, occasionally called Reid's line. It runs through two landmarks of the head that can be readily seen from the side: one is the ear opening (the tragus in the ear canal, to be exact), and the other landmark is the external junction of the eyelids—see Figure 8.2. (The so-called Frankfurt line, used in older texts, is more difficult to determine. It is pitched about 10° off the EE line.)

The EE line allows us to define two angles (in the lateral view) as shown in Figure 8.2: the pitch P of the head against the horizon and the pitch LOSA of the LOS against EE. This tells us the following:

- *How the head is held*: The angle P between the EE line compared to the horizon (or to the neck or trunk) describes the relative posture of the head. The head is said to be held erect, or upright, when the angle P between the EE line and the horizon is about 15°.
- *How the LOS is angled*: The angle LOSA between the LOS and the EE line describes how we look at a visual target: down, forward, or up.

Visual Field

The visual field is the area, measured in degrees, within which we can see objects. To the sides, each eye can see a bit over 90°, but colors are apparent only within about 65°. Upward, the visual field extends through about 55°, with colors evident only to about 30°. Downward vision is limited to about 70°, with colors visible to about 40°.

Eyeball rotation increases the visual area to the "field of fixation"; it adds about 70° to the outside, but nothing in the upward, downward, and inside directions because the eyebrows, cheeks, and nose stay in the way.

If we move our head in addition to the eyeballs, we can see nearly everything in the environment when we turn in that direction, assuming there is no occlusion by the body or other structures. Of course, mobility of the head, achieved by neck muscles, is sharply reduced in individuals with a "stiff neck," a condition often experienced by the elderly. To accommodate those suffering from less mobile neck, it is especially important to determine their naturally chosen LOS and then to place visual targets close to it.

The naturally chosen LOS to a target located 1 m or less away from the eyes is, for most people,

- Straight ahead, neither to the left or right
- Between 25° and 65° below the EE line (These numbers derive from an average LOSA of 45°, ±1.65 times the standard deviation of 12°. For more details, see Kroemer et al., 2001.)

In case we want to describe the best height of the visual target in reference to eye height, we must first consider the posture of the head. If the person holds his or her head erect (P in Figure 8.2 is about 15°), the target should be distinctly below the eyes. If he or she leans back in a seat (i.e., as P increases), the target can be placed higher: the more we lean back, the higher it can be. (This is because the LOS, as well as the EE line, moves with the posture of the head, as mentioned earlier.)

Notice that these recommendations are not "hard numbers" because different people prefer various arrangements—any one individual may not be suited for a setup that others find appropriate.

Visual Fatigue

People doing close visual work, as when working with a computer display, often complain of eye discomfort, visual fatigue, or eyestrain. While individuals' problems vary, their complaints often seem related to focusing on objects that are too far away. In other words, they are straining their eyes because their visual targets are located at a distance that differs from the minimal resting distance of accommodation, which is about 1 m away from the pupils. As we lower our angle of gaze, or tilt our head down, the resting point gets naturally closer to the eyes, to about 80 cm for a LOSA of 60° downward direction. However, the resting distance increases as we elevate the direction of sight, on average to about 140 cm at a 15° upward direction. Once again, there are large variations among people in these respects. Every person should be allowed and encouraged to adjust visual targets to her or his preferred "personal" distance and height.

Many computer users' complaints about eye fatigue are related to inadvisable placement of monitor, source documents, or other visual targets (discussed in Chapter 6) or to improper lighting conditions at their workplace, as discussed later in this chapter.

Measurement of Light (Photometry)

Physical measurements of light energy do not completely match human perceptions, meaning that subjective descriptors such as "bright" or "dim" are not solidly related to physical measurements of illuminance or luminance. The optical conditions of the human eye, the sensory perception of the stimuli, and our brain's processing modify the physical conditions of light. To account for this, the Commission Internationale de l'Eclairage (CIE) developed the "standard luminous efficiency function for photometry" in 1924 that used a model of the visual traits of the human observer. With some modifications in 1951 and 1978, the CIE (1978) standard is still valid (Rea, 2005).

The "radiant flux" measured in watt (W) describes the total energy consumed for lighting and, over time, the amount shown on the electricity bill. This translates (with a few twists—see the radiometric and photometric literature for details) into the overall "illumination" of the office, measured in lux (lx). But "luminance," measured in candela per square meter (cd/m^2), is the most important phenomenon for human vision: this is the light energy emitted by or reflected from a surface. Luminance is what enables us to see a wall, furniture, a written document, or the computer display.

Luminance of an object is determined by its incident illuminance and by its reflectance:

$$\text{Luminance} = \text{Illuminance} * \text{Reflectance} * \eth^{-1} \qquad (8.1)$$

The numerical value for illuminance is in lx and for luminance in cd/m^2. The factor $\eth^{-1}$ is omitted when the following nonmetric units are used: luminance in footlambert (fL) and illuminance in footcandle (fc). Reflectance is the ratio of reflected light to received light, stated as a percentage.

Recommendations such as those by the International Commission on Illumination (CIE) or by IESNA and ANSI in the United States for overall office illumination range from 500 to 1000 lx, even more if there are many dark (light absorbing) surfaces in the room. But if light-emitting displays are present, such as cathode ray tubes (CRTs) on mostly older computers, we can lower the overall illumination to between 200 and 500 lx to avoid degrading the image quality. In rooms with older flat-panel light-reflective displays, illumination of 300–750 lx is appropriate. However, with display technology changing rapidly, these recommendations will need to be updated frequently (Sheedy, 2006).

HOW DOES LIGHTING HELP OUR VISION?

We can see an object only if it sends light toward our eyes. We see details well if there is strong luminance contrast between the visual target and its background. Luminance is reflected illumination of light that originally came from the sun or an artificial light source (lamp, also called luminaire). If an intense light is reflected into our eyes, we experience indirect glare; if an intense light source shines straight into the eyes, we suffer direct glare. Either type of glare can make it difficult to see what we want to see.

Your accountant might appreciate your use of direct lighting (when rays from the source fall directly on the work area) because it is most efficient in terms of illuminance gain per unit of consumed electrical power, but direct light can produce high

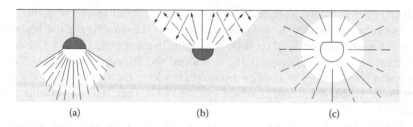

(a)   (b)   (c)

FIGURE 8.3   Lamps for (a) direct, (b) indirect, and (c) diffuse lighting. (From Morgan, C.T. et al., *Human Engineering Guide to Equipment Design*, McGraw-Hill, New York, 1963.)

glare, poor contrast, and deep shadows. The alternative is to use indirect lighting, where the rays from the light source are reflected in many different directions (diffused) at another surface, often the ceiling of the office, before they reach the work area. This helps to provide an even illumination without shadows or glare, but is less efficient in terms of use of electrical power. A compromise is to use a large translucent bowl that encloses the light source so that the light is scattered as it is emitted from the bowl's surface. This can cause some glare and shadows but is usually more efficient in terms of electrical power usage than indirect lighting. Figure 8.3 shows these types of room lighting.

In computerized offices, the overall recommended illumination level is fairly low (200–500 lx, see above) to maintain optimal luminance of the electronic displays. This illuminance may be a bit dim, especially for such tasks as reading text on a paper, so you may find it helpful to turn on a "task light" directed at the visual target that generates more luminance there without appreciably raising the overall lighting level in the office. Be careful, however, to avoid any glare that is generated by shining that light—directly or by reflection—into your eyes.

Vision Myths...Debunked*

Myth 1: Working in dim or glaring light, reading fine print, wearing glasses with the wrong prescription, or staring at a computer screen will damage your eyes.

Truth: Subjecting your eyes to any of those conditions over time can cause eyestrain, since the eye muscles struggle to maintain a clear or unwavering focus. In addition, prolonged staring can dry the front of the eye somewhat, because it reduces blinking, which helps lubricate the cornea. But fatigue and minor dryness, no matter how uncomfortable, will not permanently harm your vision.

Of course, it makes sense to minimize discomfort and eyestrain by using the following techniques:

- "Lighten up." Age tends to cloud the lens of the eye and shrink the pupil, sharply increasing the need for luminance. So if reading strains your

---

* Adapted from Kroemer et al. (2001).

eyes, consider installing brighter lights or at least moving the reading lamp closer to the page. If you are using a handheld reading device, adjust the lighting to fit you best.

- Cut the glare. Position the reading lamp so that light shines from over your shoulder, but make sure it does not reflect into your eyes from the computer monitor. Avoid reading or doing computer work near an unshaded window. And wear sunglasses if you are reading outside.
- Stop and blink. When you are working at the computer or reading, pause frequently—say, every 15 minutes—to close your eyes, or gaze away from the screen or page, and blink repeatedly. Every hour or so, get up and take a longer break.
- Find and maintain the right distance. Keep your eyes at the same distance from the screen as you would from a book. If that is uncomfortable, buy a pair of glasses with a prescription designed for computer work or use "progressive-addition" bifocals, which have gradually changing power from the top to the bottom of the lens.
- Lower the screen. Keep the top of the screen below eye level. Gazing upward can strain muscles in the eye and neck.
- Use a document holder. Put a support for reading materials next to the screen, at the same distance from your eyes.
- Clean your screen and your glasses. Dust and grime can blur the images.
- If these measures do not reduce the strain from either reading or computer work, have an optometrist or ophthalmologist check whether you need to start wearing glasses or to have your prescription changed.

Myth 2: The more you rely on your eyeglasses or contact lenses, the faster your eyesight deteriorates.

Truth: This myth is based on the misconception that artificial lenses do your eyes' work for them, which supposedly tempts your eyes to get lazy and weak. However, artificial lenses merely compensate for a structural defect of the eye—an improperly shaped eyeball or an excessively stiff lens—that prevents proper focusing, despite the best efforts of the lens muscles. When you wear appropriate glasses or contact lenses, your eyes avoid excessive efforts in tasking the eye muscles to overcome the vision defect. The corrective lenses allow them to work just as hard as the muscles in a normal eye.

Myth 3: Eye exercises can help many people see better.

Truth: Eye exercises can help some children whose eyes have major binocularity problems, such as crossing, misalignment, or inability to converge. But claims that the exercises can help many children read better are unsubstantiated. Some people have asserted that eye exercises can help us not only read better but also sharpen visual acuity, boost athletic performance, and help correct numerous problems in both children and adults. The truth is such claims are unsupported and implausible.

COLOR PERCEPTION

Interestingly, human color perception—in spite of hundreds of years of research and study—is still not completely understood. What we do know is that vision involves the nearly simultaneous interaction of the two eyes and the brain through a network of neurons, receptors, and other specialized cells. Sunlight contains all visible wavelengths, but objects struck by the sun's rays absorb some radiation. These objects then reflect that light themselves, and the light that they transmit has a different energy distribution than what they received.

When we look at a particular object, one that is transmitting light, our eyes do not analyze the spectral composition of the light. Instead, our brain simply "classifies" incoming signals from different groups of wavelengths to label them "colors" by experience. Color is a subjective experience; in fact, two objects that we perceive as identical in color may actually have different spectral contents. The same spectral content easily gets called a different "color name" by several individuals. Human color perception, then, is a psychological experience, not a single specific property of the electromagnetic energy of light that we see.

Several theories and concepts attempt to explain human color perception. One concept of equivalent-appearing stimuli underlies a system of color measurement and specification called colorimetry. We humans can perceive the same color even if it consists of various mixes of the three primary colors, red, green, and blue (the Young and Helmholtz theory). Therefore, human color vision is called trichromatic, and the often used CIE chromaticity diagram plots colors by the amounts of standard red, blue, and green primaries that match them (Boyce, 2014). In addition to the Young and Helmholtz theory, several other concepts try to explain human color vision; we recommend looking up Hering's opponent colors theory and Judd's zone or stage theories and watching for new approaches.

AESTHETICS AND PSYCHOLOGY OF COLOR

While the physics of color stimuli arriving at the eye can be well described, perception, interpretation, and reaction to colors are highly individual, nonstandard, and variable. A person's judgment of the perceived color of a visual stimulus depends on subjective impressions experienced when viewing the stimulus, and the judgment varies with the viewing conditions and the kind of stimuli.

People believe in and describe emotional reactions to color stimuli. For example, reds, oranges, and yellows are usually considered "warm" and stimulating. Violets, blues, and greens are often felt to be "cool" and to generate sensations of cleanliness and restfulness. However, likes or dislikes of certain colors and their combinations are regionally very different; travel to Asia and Europe, for example, makes this apparent. Although experimental evidence on these effects is missing (Post, 1997) or controversial (Kwallek and Lewis, 1990), color schemes are often applied to work and living areas to achieve emotional responses. This is why, in Western countries, your bank probably uses blues (denotes "security"), your grocer greens ("freshness"), your church deep shades of purples and reds ("spirituality").

COMPUTER DISPLAY

Older displays used a CRT to generate a picture that the human eye could perceive. Newer technological devices have flatter screens, utilizing advanced light-reflective and light-emissive displays, passive or active thin-film transistor matrices. In the United States, the 2008 ANSI/HFES 100-2007 standard describes optical properties and contains guidance for appropriate human engineering.

So far, it has generally turned out that whatever the technology in practice, the optical properties of the display determine the suitable lighting of the room in which the monitor is located, as stated earlier. As of this writing, in December 2015, the most-used technology relies on CRTs with liquid crystal displays and different sources of backlighting. Nearly all of them can be used in brightly lit environments. However, emerging technologies quickly spawn new ways to make electronically generated events visible to the eye, which can change the illuminating conditions of the environment. Therefore, even current recommendations for example by the Illuminating Engineers Society and in the most recent ANSI/HFES 100 or in OSHA Document 3092 can quickly become outdated. Consequently, the human factors specialist should follow technical developments closely to stay up to date.

PLACING THE DISPLAY

The eye "sees" what arrives at the retina—the incoming light's energies and wavelengths are distributed over the area of stimulated rods and cones. Ideally, the arriving image solely represents the electronically generated display, but of course, it can contain the reflections of ambient light sources that are mirrored at the surface of the display. If the mirrored light is strong enough, we speak of glare. If it is too strong, it diminishes the apparent contrast, or color, and hence washes out the original image so that we may find it difficult, if not impossible, to discern it anymore.

Avoiding Indirect Glare

An example of distracting glare is when you are driving at night and the headlights of the car behind yours are reflected in your rear view mirror, striking your eyes and reducing your ability to see the road ahead. In the office, your view of text or other images in the computer display may be disturbed by reflected glare (perhaps without you being consciously aware of it). Examples of indirect glare are shown in Figure 8.4; here, the sources of the glare are from a lamp and the sun. An imperfect way to reduce glare is to apply some sort of treatment to the front of the reflecting surface of your monitor. A filter is commonly used that absorbs some energy of the incoming light, often in certain wavelengths (colors), but it also absorbs energy from the emitted image, which—unfortunately—reduces the luminance contrast that arrives at the eye. A related approach is to use micromesh or microlouvers that limit the directions from which stray light may fall onto the monitor surface. The disadvantage of the latter technique is that the observer must position his or her eyes in exactly the proportionate position, forcing head and upper body into a proscribed and immutable posture. Alternately, we can use coatings on the surface or roughen the surface to affect the reflections of incoming light—but the basic problem with all of

FIGURE 8.4 Indirect glare on the surface of a computer display can be caused by a lamp or the sun or a window.

these measures is that they treat the symptoms, the reflections—instead of eliminating the source of the disturbance in the first place.

As with many things in life, it is much better to tackle the origin of the problem rather than treat the effects. Some guidelines: avoid wearing a white blouse or shirt that would be mirrored in the monitor. Move or turn off a lamp if its light is reflected into your eyes, or put a shield between the luminaire and the monitor. And if it is a window whose bright surface appears on your display, then draw a (dark) curtain or window treatments in front of it. Of course, you often can simply reposition your workstation together with the monitor so that the bright lamp or window is to your side or somewhere else where it does not get reflected into your eyes. The same solutions work if a bright light is mirrored from your work desk, or a sheet of paper on which you try to write, as depicted in Figure 8.5. While you are at it, you could make sure that there are no smooth, polished, "shiny" surfaces at all in your field of view that can act as mirrors to disturb your eyesight.

Avoiding Direct Glare

Let us go back to your car for a moment. Those reflected headlights gave you indirect glare; now, imagine a car coming toward yours at night with high beams fully on. The headlights of the approaching car can "blind" you for a moment, if the ample light energy shining directly on your retina overpowers the subtle image of the road that you saw at low-level luminance. In the office, it is probably a task light pointed at you or immense light flooding in from a window in front of you—as depicted in Figure 8.6—that optically overpowers the image presented on the

FIGURE 8.5 Indirect glare caused by light that is reflected on a shiny work surface.

FIGURE 8.6 An example of direct glare. Bright light from a window shines into the eyes where its energy overpowers the weaker image coming from a computer display.

monitor. What makes you regain the ability to discern the details on the display are the same measures as used to eliminate indirect glare:

- Reduce the intensity of the incoming light (turn off the lamp, draw a curtain across the window).
- Reposition yourself, your monitor, and your entire workstation by about 90°.

The best ways to provide glare-free lighting from windows and luminaires are to position them strategically. Locating light sources to the left and right sides of the operator, or overhead, is not likely to cause indirect glare, assuming that there are no reflecting ("shiny") surfaces at the workstation.

- Do not place a lamp or bright window in front of the person because this would cause direct glare.
- To avoid indirect glare, avoid placing a light source behind the person because the light would be mirrored on the display surface.

In summary, the optical properties of the display and the fact that you want to clearly discern what appears on the screen make placement of the monitor in specific relations to your eyes necessary. The ideal placement is as follows: in front of the observer, at that person's personal reading distance (usually about half a meter) from the eyes, and at such height that the preferred LOS meets the center of the display.

The screen should be about perpendicular to the LOS. The LOS should be distinctly tilted below the EE line, exactly as the person who will be operating the display prefers it. We note that current laptop computers naturally come closer to this recommendation than most desktop computers.

With the advent of wireless and handheld computers and technological advances in input devices, avoiding glare should be relatively easy to do. Choose your light sources and place them carefully and strategically.

ERGONOMIC DESIGN RECOMMENDATIONS

The characteristics of human vision discussed earlier provide important background information that you should use when designing your work environment for proper vision. The most important lighting concepts are the following:

- Proper vision requires sufficient quantity and quality of lighting, carefully arranged to provide luminance of visual targets while avoiding glare.
- What mostly counts is the luminance of an object—the light reflected or emitted from it—which meets the eye.
- The acuity of seeing an object is greatly influenced by strong luminance contrast between the object and its background, including shadows.
- Avoid unwanted or excessive glare. Direct glare meets the eye straight from a light source, such as the sun or a lamp shining into your eyes. Indirect glare is reflected from a surface into your eyes, such as the sun or a lamp mirrored in your computer screen.
- Using colors for the visual target can be helpful if the colors are well chosen, but color vision requires sufficient light, and luminance contrast is more important than coloring.
- The colors of the various surfaces and spaces in the workspace affect the reflectance and hence luminance of these areas and may affect mood and attitudes of the people in it.
- Place the display of your computer "close and low," directly behind the keyboard.

Optimal lighting conditions for good vision depend on many factors, including the task at hand, the objects that need to be seen, and the individual's eye health and conditions.

- Overall recommendations for office illumination range from 500 to 1000 lx, even more if there are many dark (light absorbing) surfaces in the room.
- If light-emitting displays are present, such as screens on computers, the overall illumination should be between 200 and 500 lx.
- In rooms with older flat-panel light-reflective displays, illumination of 300–750 lx is suitable.

ERGONOMIC GUIDELINES

We suggest that you follow valid recommendation for the overall illumination level in the office. In the United States, guidelines can be found in the current CIE and IESNA *Lighting Handbooks*, in OSHA Document 3092 in documentation, by the Illuminating Engineers Society and in the most recent ANSI/HFES 100.

When possible, select indirect lighting, where all light is reflected at a suitable surface (at the ceiling or walls of a room, or within the luminaire) before it reaches the work area. This helps to avoid direct and indirect glare.

Use several low-intensity lights instead of one intense source, placed away from the LOS; this avoids direct glare.

Place any existing high-intensity light sources (including windows) outside a cone-shaped range of 60° around the LOS; this avoids direct glare.

Shine a task light on your visual target if the overall illumination is too dim to generate sufficient luminance.

Properly distribute light over the work area, which should have dull, matte, or other nonpolished surfaces; this avoids indirect glare.

The naturally chosen LOS to a target not more than 1 m away from the eyes is

- Straight ahead, not veering to the left or right
- Between 25° and 65° below the EE line (If the person holds the head straight, the target should be distinctly below the eyes, never higher.)

Please keep in mind that these recommendations are not "hard numbers" because different people prefer various arrangements—what is good for one individual may not suit everybody else.

REFERENCES

Boyce, P. R. (2014). *Human Factors in Lighting* (3rd ed.). Boca Raton, FL: CRC.
CIE (Commission Internationale de l'Eclairage). (1978). *Light as a True Visual Quantity: Principles of Measurement.* Vienna, Austria: CIE, CIE Publication 41.
Delleman, N. J. (2000). Evaluation of head and neck postures. In *Proceedings of the XIVth Triennial Congress of the International Ergonomics Association and 44th Annual Meeting of the Human Factors and Ergonomics Society.* Santa Monica, CA: Human Factors and Ergonomics Society, pp. 5-732–5-735.

Kroemer, K. H. E., Kroemer, H. B., and Kroemer-Elbert, K. E. (2001). *Ergonomics: How to Design for Ease and Efficiency* (2nd ed.). Upper Saddle River, NJ: Prentice Hall /Pearson.

Kwallek, N. and Lewis, C. M. (1990). Effects of environmental colour on males and females: A red or white or green office. *Applied Ergonomics* 21, 275–277.

Morgan, C. T., Cook, J. S., Chapanis, A., and Lund, M. W. (1963), *Human Engineering Guide to Equipment Design.* New York: McGraw-Hill.

Post, D. L. (1997). Chapter 25: Color and human–computer interaction. In M. Helander, T. K. Landauer, and P. Prabhu (Eds.). *Handbook of Human–Computer Interaction* (2nd ed.) Amsterdam, the Netherlands: Elsevier, pp. 573–615.

Rea, M. S. (2005). Chapter 68: Photometric characterization of the luminous environment. In N. Stanton, A. Hedge, K. Brookhuis, E. Salas, and H. Hendrick (Eds.). *Handbook of Human Factors and Ergonomics Methods.* Boca Raton, FL: CRC.

Sheedy, J. (2006). Vision and work. In W. S. Marras and K. Karwowski (Eds.). *The Occupational Ergonomics Handbook* (2nd ed.). Boca Raton, FL: CRC.

9 Hearing and Sound

I think there's something strangely musical about noise.

—**Trent Reznor, musician, "Nine Inch Nails"**

You're short on ears and long on mouth.

—**John Wayne**

OVERVIEW

Most humans are "social creatures": we want to know the people in the office around us, get along with them, and talk and work with them. (Read more about social interactions in the office in Chapter 2.) Speaking and listening are, in fact, essential work tools for most of us. However, we also want some privacy, and we certainly do not want to be disturbed or disrupted by the sounds that our colleagues make or by noises coming from phones and other office equipment, from machinery, or from outside the office. If loud sounds abound, they can make us uncomfortable—even decrease our productivity—and may reduce our hearing ability in the longer term.

PSYCHOACOUSTICS

Psychoacoustics describes how we perceive sounds in relation to their physical properties. This topic will be covered in some detail in the following text.

We are comfortable when the acoustical engineer succeeds in designing an office environment that

- Transmits the sounds we want to hear and does so reliably
- Is pleasant to us with respect to the overall "background noise"
- Minimizes sound-related annoyance and stress
- Minimizes disruption of speech communications
- Prevents hearing loss

Note: You may skip the following part and go directly to the "Ergonomic Design Recommendations" section at the end of this chapter—or you can get detailed background information by reading the following text.

HOW EXACTLY DO WE HEAR?

Sound can reach our ears via two different paths. Airborne sound travels through the ear canal and excites the eardrum and the structures behind it, as described in the following text. Sound may also be transmitted through bony structures in our head, but this requires much higher intensities to be similarly effective. In this text, we cover airborne sound.

How well we hear—how well we distinguish sounds and understand their meaning—depends on the intensity and frequency of the sounds at their source, how they are transmitted to us (which ideally occurs without interference by other sounds), and how healthy or "sound" our individual hearing organs are.

ACOUSTICS

Acoustics is the science and technology of sound—its production, transmission, and effects. It describes the structure of sound by its physical characteristics, frequency, amplitude, and duration. Psychoacoustics establishes relations between the physics of sound and our individual perception. It uses descriptors such as pitch, timbre, loudness, noise, and understanding of speech.

Let us start with the definition of "sound": we define it as any vibration—passage of zones of compression and rarefaction—through the air or any other physical medium that stimulates an auditory sensation. Sound is first generated and then transmitted to us; thereafter, we hear it, understand it, and like it (or not). This simple source–path–receiver concept tells the acoustic engineer to manipulate sound at its source—speak a message clearly and pleasantly, for example—or prevent generation of a noise. We can also manipulate the transmission of a sound by using high-fidelity speakers if we want the signal to remain clear, for example, or we can reduce the energy propagation of annoying sounds by using materials that dam or absorb sound along its path. An example of the latter may involve sound-absorbing materials like drapery or acoustic tiles.

SOUND, EAR ANATOMY, AND HEARING

Figure 9.1 is a sketch of the ear. Sound waves arriving from the outside are collected by the outer ear (auricle or pinna) and funneled into the auditory canal (meatus) to the eardrum (tympanic membrane), which vibrates according to the frequency and intensity of the arriving sound wave. Resonance effects of the auricle and meatus amplify the sound intensity by 10–15 decibels or dB (see below for an explanation of the dB unit) when it reaches the eardrum.

In the middle ear, the sound that arrived via the eardrum is mechanically transmitted by three ear bones, called the ossicles, to the oval window. Those three ear bones are the hammer (malleus), the anvil (incus), and the stirrup (stapes).

Mechanical properties of the ear bones, and the fact that the area of the eardrum is much larger than the surface of the oval window, increase the intensity of the sound coming from the eardrum about 15 times when it arrives at the oval window.

Both the outer and the middle ears are filled with air. The Eustachian tube connects with the pharynx, which allows the air pressure in the middle ear to remain equal to external pressure. While air fills the outer and middle ear, the inner ear is filled with a watery fluid, called endolymph or perilymph, which does not need pressure adjustment.

A "clogged" Eustachian tube can delay the equalization of pressure between the middle ear and the environs. In a rapidly climbing or descending airplane, when you feel ear

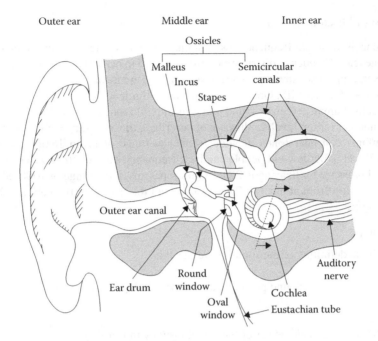

FIGURE 9.1 Schematic of the human ear. (Adapted from Kroemer, K.H.E., *Fitting the Human*, CRC, Boca Raton, FL, 2016.)

pressure, feel pain, or cannot hear well, you may try to open the tube by chewing gum or by willful excessive yawning but "pumping" your outer ears with your hands will not help your middle ears because the ear drums separate them tightly.

The inner ear contains the cochlea, a bone canal shaped like a snail shell with about two and one-half turns. Sound waves move through the cochlea as fluid shifts from the oval window to the round window. The motion of the fluid distorts the basilar membrane that runs along the cochlea. This in turn stimulates sensory hair cells (cilia) in the organs of Corti, which are located on the basilar membrane. The different sensors respond to specific frequencies and generate impulses that are then transmitted along the auditory (cochlear) nerve to the brain for interpretation.

Jacqueline is flying to visit her brother in Paris, and she is nervous. It is the first time she has flown in an airplane with her baby, who just turned 4 months old. She has flown plenty of times and has all too often experienced the misery of being trapped in an airplane close to a fussy screaming baby, with frantic parents trying to calm the offending infant, and she does not want to subject her fellow travelers to that experience. Her brother has three children, and he advised her to give the baby a bottle during takeoff and landing. He said that when the baby swallows liquid, it will relieve the pressure on the ears, and crying will not be an issue. Jacqueline sighs as she packs up the diaper bag in preparation for her trip. She only hopes that trick will work; otherwise, it will be one long airplane ride.

HUMAN HEARING RANGE

A tone is a single-frequency oscillation, while a sound contains a mixture of frequencies. Frequency distributions are measured in hertz (Hz), and their intensities (power, amplitude, sound pressure levels [SPLs]) are measured in logarithmic units known as decibels (dB). One reason that a logarithmic scale is used for measurement is that the human perceives sound pressure amplitudes in a roughly logarithmic manner; another reason involves the ease of describing the wide range of human hearing.

Infants can hear tones in wide frequency range of up to 16,000–20,000 Hz (16–20 kHz), but few old people can hear frequencies above 10 kHz. The minimal sound pressure for hearing is 20 μPa (micropascal) in the range of 1000–5000 Hz. We feel pain from about 140 Pa on and cannot tolerate more than about 200 Pa—see Figure 9.2.

Decibel

The basic definition of the sound power level is

$$\text{Sound power level (in dB)} = 10 \log_{10} (Pw/Pw_0)$$

where
    Pw is the acoustic power of the sound (usually in watts)
    Pw_0 is the power at a reference level, usually set to 10^{-12} W, which is the hearing threshold of a young healthy ear

The decibel or dB (meaning 1/10th of a bel; *bel* is named after Alexander Graham Bell) is the unit name given to any such ratio.

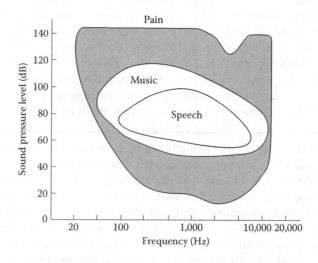

FIGURE 9.2 Ranges of adult human hearing. (Adapted from Kroemer, K.H.E., *Fitting the Human*, CRC, Boca Raton, FL, 2016.)

The SPL is also commonly measured in dB. It is the ratio between two sound pressures, of which one, the threshold of hearing (P_o), is used as reference. Since power is proportional to the square of the pressure, the equation is

$$SPL = 10\log\left(\frac{P^2}{P_o^2}\right) \text{ or } 20\log_{10}\left(\frac{P}{P_o}\right) \text{in dB}$$

where
P is the root-mean-square sound pressure for the existing sound
P_o is the threshold of hearing, the reference

With these values, the dynamic range of human hearing from 20×10^{-6} to 200 Pa is

$$20\log_{10}[200/(20 \times 10^{-6})] = 140 \text{ dB}$$

Ranges of SPLs are shown in Figure 9.2.

PSYCHOPHYSICS OF HEARING

While physical measurements (such as in dB) can explain acoustical events, people interpret them and react to them in very subjective manners—you may find certain sounds loud versus soft, for example, or pleasant rather than noisy, but your neighbor or colleague may vigorously disagree with your interpretation. The "loudness" of a tone (or complex sound) that a given individual perceives depends on both its intensity and its frequency; put differently, the subjective experience of combined frequency and amplitude of a sound is loudness. Compared to the intensity at 1000 Hz, at lower frequencies the SPL must be increased to generate the feeling of "equal loudness." For example, the intensity of a 50 Hz tone must be nearly 85 dB to sound as loud as a 1000 Hz tone with about 50 dB. However, at frequencies in the range of about 2000–6000 Hz, the intensity can be lowered and the sound still appears as loud as at 1000 Hz. Above about 8000 Hz, the intensity must be increased again, even above the level at 1000 Hz, to sound equally loud. The "equal-loudness contours" (called "phon" curves) are shown in Figure 9.3.

These perceptions of equal loudness indicate that there are nonlinear relationships between "pitch" (the perception of frequency) and "loudness" (the perception of intensity). *Timbre* is even more complex because it depends on changes in frequency and intensity over time.

Filters that are applied to sound-measuring equipment imitate the differences in human sensitivity to tones of different frequencies. These filters "correct" the physical readings to reflect what an individual perceives. Different filters have been used, identified by the first letters of the alphabet. The "A" filter is most often used because it is close to an individual's response at 40 dB. A-corrected decibel values are identified by the notation dBA or dB(A).

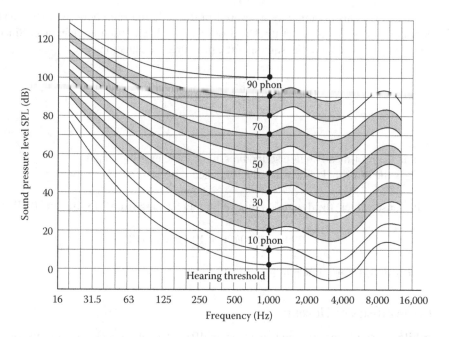

FIGURE 9.3 Phon curves are lines along which combinations of sound pressure level and frequency are perceived as having the same "loudness" as at 1000 Hz. (Adapted from Kroemer, K.H.E., *Fitting the Human*, CRC, Boca Raton, FL, 2016.)

ACOUSTIC FACTS

We conclude this section on acoustics with some additional facts about how we humans hear:

- *Directional hearing*—Individuals can determine where a sound is coming from by using the difference in arrival times or intensities at their ears.
- *Distance hearing*—The ability to determine the distance of a source of sound is related to the fact that sound energy diminishes with the square of the distance traveled, but the human perception of energy depends also on the frequency of the sounds, as just discussed. Consequently, a sound's source appears more distant when it is low in intensity and frequency and appears closer when it is high in intensity and frequency.
- *Difference and summation tones*—Two tones that are sufficiently separated in frequency are perceived as two distant tones. When two such tones are very loud, we may hear two supplementary tones. The more distinct tone is at the frequency difference between the two tones; the quieter tone is the summation of the original frequencies. For example, two original tones, at 400 and 600 Hz, generate a difference tone at 200 Hz and a summation tone at 1000 Hz.
- *Common-difference tone*—When several tones are separated by a common frequency interval, we hear an additional frequency based on the common

difference. This effect explains how we may hear a deep bass tone from a sound system that is physically incapable of emitting such a low tone.

- *Aural harmonics*—We may perceive a pure sound as a complex one because the ear can generate harmonics within itself. These "subjective overtones" are more pronounced with low-frequency tones than with high-frequency ones, especially if these are about 50 dB above the threshold for human hearing.
- *Intertone beat*—If two tones differ only slightly in their frequencies, the ear hears only one frequency, called the "intertone"; this is halfway between the frequencies of the original tones. The two tones are in phase at one moment and out of phase at the next, causing the intensity to wax and wane; consequently, we hear a beat.
- *Doppler effect*—As the distance between the source of sound and the ear decreases, we hear an increasingly higher frequency; as the distance increases, the sound appears lower. The larger the relative velocity, the more pronounced the shift in frequency. The Doppler effect can be used to measure the velocity at which source and receiver move against each other.
- *Concurrent tones*—When two tones of the same frequency are played in phase, they are heard as a single tone, with loudness being the sum of the two tones. Two identical tones exactly opposite in phase cancel each other completely and cannot be heard. This physical phenomenon (called destructive interference or phase cancellation) can be used to suppress the propagation of acoustical or mechanical vibrations.

HEARING LOSS

Hearing loss is the third most common physical ailment after arthritis and heart disease, so it is an important and far-reaching public health concern. Hearing loss may be a sudden or a gradual decrease in how well you can hear, and because we cannot see hearing loss, it may go undetected for a period of time, during which the afflicted individual may be perceived as aloof, confused, or uncooperative.

Using the World Health Organization's definition for hearing loss—not being able to hear sounds of 25 dB or less in the speech frequencies—research shows that overall, about 30 million people in the United States alone, nearly 13% of the population, had hearing loss in both ears. That number jumps to about 48 million, more than 20%, for people who have hearing loss in at least one ear (Lin, 2011).

Degrees of hearing loss vary from mild, moderate, severe to profound, and it may be temporary or permanent. Hearing loss can be congenital, meaning you are born without hearing; it can happen gradually over time; or it can result from an injury.

In adults, the most common causes of hearing loss are (1) aging and (2) noise. Age-related hearing loss is a common occurrence which results from changes in the inner ear that happen as you grow older; the presbycusis may be mild, or it may be severe, but it is always permanent. In older people, hearing loss is sometimes confused with or misdiagnosed as dementia.

Other causes of hearing loss—which we will not cover further in this text, since they are largely unrelated to office work and affect mostly children—include earwax buildup, an object in the ear, injury to the ear or head, ear infection, a ruptured eardrum, and other conditions that affect the middle or inner ear.

NOISE-INDUCED HEARING LOSS

Noise-induced hearing loss (NIHL) may happen suddenly or slowly, over time. Sudden NIHL results from exposure to short-duration sound of high intensity, like the sound of an explosion. This sound can damage any or all of the structures of the ear, in particular the hair cells—the cilia—in the organ of Corti, which may be torn apart. The result is immediate, severe, and permanent hearing loss. In fact, sudden NIHL from explosions and gunfire is one of the most common injuries in combat troops. Often, severe ringing in the ears, known as tinnitus, accompanies the hearing loss, and this exhausting and unrelenting condition can be as debilitating as the hearing loss itself.

Gradual NIHL results from exposure to everyday noises, such as listening to very loud music, having a noisy work environment, or using construction equipment like a jackhammer without adequate ear protection, over long periods of time. The amount and severity of hearing loss depends on sound levels. Sound levels of less than about 100 dB often cause a short-term hearing loss, measured as temporary threshold shift (TTS). During quiet periods, hearing returns to its normal level.

Repeated exposures to sounds that cause TTS may gradually bring about a permanent threshold shift (PTS), just as a loud explosion can do (more information on this is found in the "Noise in the Office" section). Often, the damage is confined to a specific area on the cilia bed on the cochlea, related to the frequency of the sound. With continued sound exposure, more hair cells are damaged, which the body cannot replace; also, nerve fibers that communicate signals from that region to the brain may degenerate, accompanied by corresponding impairment within the central nervous system.

Sound level, frequency, and duration of exposure (singly or repeatedly) are the critical descriptors for sounds that can damage hearing. Sound levels below 75 dBA apparently do not produce permanent hearing loss, even at about 4000 Hz, where people are particularly sensitive. At higher intensities, however, the amount of hearing loss is directly related to sound level and its duration. In the United States, current OSHA regulations allow 16 hours of exposure to 85 dBA, 8 hours to 90 dBA, 4 hours to 95 dBA, etc. Other countries have different regulations and limits in terms of allowable exposure to noise. To illustrate, in Europe, 8 hours at 90 dBA are allowed, but only 4 hours at 93 dBA, or 16 hours at 87 dBA.

Ranges of noise, lengths of exposure, and resulting effects are shown in Table 9.1.

TABLE 9.1
Decibel Levels and Consequences

Decibel Level	Length of Exposure	Examples	Level of Pain
Below 85	Exposure over any amount of time is safe.	Normal conversation; a babbling brook	No pain
85–90	Exposure to 90 dB over long period of time may lead to hearing loss.	Passing diesel truck; lawn mower	No pain
90–100	Hearing loss occurs over a length of time.	Handheld drill; food processor	Discomfort
100–130	Exposure over a short period of time causes hearing loss.	Rock concert; sandblasting; jackhammer	Tinnitus possible after exposure; discomfort with pain above 120 dB
140+	Immediate hearing loss after single exposure.	Explosion; 12-gauge shotgun blast	Pain

Simple subjective experiences can indicate the existence of hazardous sound exposure. Indications of a dangerous sound environment are

- Hearing a sound that is appreciably louder than conversational level
- Hearing a sound that makes it difficult to communicate
- Experiencing the sensation of ringing in the ear (tinnitus) after having been in the sound environment
- Experiencing the sensation that sounds seem muffled after leaving the noisy area

Permanent hearing loss ranges from mild to profound. When we know the cause and nature of the hearing loss, we can work to prevent it, or if it has already occurred, we can attempt to alleviate it through hearing aids. Of course if the underlying nervous structure is damaged, the hearing loss is generally not treatable under the auspices of current technology and expertise.

NIHL increases most rapidly in the first years of exposure; after many years, it levels off in the high frequencies, but continues to worsen in the low frequencies (National Institutes of Health, 1990).

In Western countries, with their specific noises, NIHL usually starts in the range of 3000–6000 Hz, particularly around 4000 Hz. Then, it extends into higher frequencies, culminating at about 8000 Hz. However, reduced hearing at (and above) 8000 Hz is also brought on by aging. This may make it difficult to distinguish between environment- and age-related causes of NIHL.

AUDIOMETRY

Assessment of hearing ability, especially its loss, includes measures of the auditory thresholds (sensitivity) at various frequencies. Pure-tone audiometry, done by ophthalmologists and optometrists, is often combined with measures of understanding speech, which is covered under 'Voice Communications' below.

IMPROVING HEARING ABILITY

Many people, especially as they get older, experience a lessening ability to hear, predominantly in the higher-frequency ranges. Devices to aid impaired hearing can offer a significant improvement in understanding speech, especially when these devices utilize electronics to amplify certain bandwidths of sound—depending on the existing hearing deficiency—and filter out ambient noise.

In reality, however, many people with impaired hearing become very frustrated when trying to use these hearing aids, leading many to skip hearing aids altogether. Fortunately, hearing aids have improved significantly over the past decade, and they will continue to evolve as technology allows them to.

A hearing aid is a small electronic device worn in or behind the ear, and it essentially makes some sounds louder. It has three basic parts—microphone, amplifier, and speaker—and works by receiving sound through a microphone, which converts the sound waves to electrical signals and sends them to an amplifier. The amplifier increases the power of the signals and then sends them to the ear through a speaker.

Hearing aids are primarily useful in improving the hearing and speech comprehension of people who have hearing loss due to aging, disease, or noise-induced injury and whose hearing loss results from damage to the small sensory hair cells in the inner ear (the cilia). This type of hearing loss is called sensorineural hearing loss. The hearing aid magnifies sound vibrations entering the ear. Surviving hair cells detect the larger vibrations and convert them into neural signals that are passed along to the brain. The greater the damage to a person's cilia, the more severe the hearing loss and the greater the hearing aid amplification needed to make up the difference. However, there are practical limits to the amount of amplification a hearing aid can provide. In addition, as mentioned earlier, if the inner ear is too damaged, even large vibrations will not be converted into neural signals. In this situation, a hearing aid would be ineffective.

Hearing aids work differently depending on the electronics used. The two main types of electronics are analog and digital:

- Analog aids convert sound waves into electrical signals, which are amplified. These hearing aids have more than one program or setting, and the settings can be changed to fit the listening environment, say from a small quiet office to a crowded football arena.
- Digital aids convert sound waves into numerical codes, similar to the binary code of a computer, before amplifying them. Because the code also includes information about a sound's pitch or loudness, the aid can be specially programmed to amplify some frequencies more than others (NIH, 2015).

Hearing aids also come in various styles, including the following:

- Behind-the-ear (BTE) aids, which consist of a hard plastic case worn behind the ear and connected to an earmold that fits inside the outer ear. The electronic parts are held in the case behind the ear. The newest BTEs are small, open-fit aids that fit behind the ear completely, with only a narrow tube inserted into the ear canal, enabling the canal to remain open.

- In-the-ear (ITE) aids, which fit completely inside the outer ear and are used for mild to severe hearing loss. Some ITE aids include a telecoil, a small magnetic coil that allows users to receive sound through the circuitry of the hearing aid rather than through its microphone. This makes it easier to hear conversations over the telephone and helps people hear in public facilities that have installed special sound systems, called induction loop systems.
- Canal aids, which fit into the ear canal, available in two styles: in-the-canal hearing aid is made to fit the size and shape of a person's ear canal; completely-in-canal hearing aid is nearly hidden in the ear canal. Both types are used for mild to moderately severe hearing loss; their reduced size limits their power and volume.

Implantable hearing aids are also available, and they work differently from hearing aids worn externally. A middle ear implant attaches to one of the bones of the middle ear and directly moves the bones. A bone-anchored hearing aid bypasses the middle ear altogether and attaches to the bone behind the ear, transmitting sound vibrations directly to the inner ear through the skull. Of course, implants require surgery, and any invasive procedure means the individual should carefully weigh the benefits versus the risk of the procedure (NIH, 2015).

Important to note is that a hearing aid will *not* restore normal hearing, so avoiding hearing loss is key. While you cannot change the passage of time and the decreased hearing ability that accompanies it, we should do what we can to minimize NIHL and protect our hearing as much as possible.

NOISE IN THE OFFICE

Noise is defined as a sound that is unwanted, objectionable, annoying, or unacceptable to a person—it is often but not always loud. Consider the following scenarios: On Monday evening, as you are settling in to sleep, you remember that you forgot to call the plumber about that dripping bathroom sink. And just as you remember this, you hear the dripping—one drop after another, 4 seconds apart, never ceasing, and unrelenting. It takes an hour to get to sleep. Tuesday evening, as you are preparing to doze off, you hear staccato noises, sharp and shrill; your neighbor is practicing his brand new electric guitar again, and he is playing in his garage with the garage door wide open. His guitar wails and whines for an hour. You suffer a second night of poor sleep. Dripping water makes single, short tones of low intensity: we consider this "noise" just as we consider the loud, lasting, complex sounds of the neighbor's music to be noise.

Noise can be measured and described in physical units, but in an office setting, the effects of noise are usually psychological and subjective, and they depend on many circumstances. Noise in the office can

- Make it difficult to hear and understand wanted sounds, especially spoken communications
- Create negative emotions, including feelings of surprise, frustration, and anger

- Interfere with a person's sensory and perceptual capabilities and hence degrade task performance
- Temporarily or permanently reduce our hearing capability if the noise has more than about 80 dB, even if one considers it pleasant such as when listening to "loud" music

Voice Communications

In an office setting, much of the exchange of information occurs via speech; we may be speaking with other people in the same room, we may be speaking on the telephone, or perhaps we are teleconferencing, skyping, or video chatting by some other means. In any event, we are using vocal correspondence to communicate, with or without the assistance of body language and visual cues.

Understanding Speech

Voice communications use frequencies from about 200 to 8000 Hz, with the lower range of about 1–3 kHz (1000–3000 Hz) most important for understanding. Men use more of the low-frequency energy than women do.

Difficulty in understanding language is often associated with problems of differentiating speech sounds in their high-frequency ranges. In many languages, consonants consist of high frequencies that have less speech energy than vowels and therefore are more difficult to understand, especially for people with NIHL and older people. In addition, other sounds such as background noise, competing voices, or reverberation may interfere with the listener's ability to receive information and to communicate. Consequently, important informational content of speech may be unclear, unusable, or inaudible.

The ability to understand the meaning of words, phrases, sentences, and whole communications is called intelligibility. This is a psychological process that depends on acoustical conditions. For satisfactory communication of most voice messages, when these messages are emitted in noise, we require at minimum 75% intelligibility. Perhaps you have at some point experienced that a discussion on the telephone was difficult to understand due to background and office noise around you, yet a conversation with your colleague just moments later was easy to comprehend, even though both dialogues appeared to be conducted at the same level of sound. Direct, face-to-face communication provides visual cues that enhance speech intelligibility, even in the presence of background noise. This is because nonverbal visual cues combine with speech to help direct us to the meaning of the messages given. Indirect voice communications, such as by voice-only telephones, lack the visual cues and thus might be less intelligible.

As a side note, this is also why when we use text messages—or e-mails, tweets, posts, or any other means of communicating that lack nonverbal cues—to communicate, we can easily misinterpret or be misinterpreted. Without the benefit of facial expressions, vocal inflections, and physical gestures, it can be difficult to interpret whether our conversation partner is being serious or sarcastic, funny or ironic, cold and formal, or simply busy. Of course, emoticons, emojis, and punctuations like exclamation points help, but they can only do so much.

Speech-to-Noise Ratio

The intensity level of a speech signal relative to the level of ambient noise fundamentally and profoundly influences the intelligibility of speech. The commonly used "speech-to-noise ratio" (S/N) is really not a fraction but a difference. Consider the example of a speech, given at 80 dBA in noise of 70 dBA; here, the S/N is simply +10 dB. With an S/N of +10 dB or higher, people with normal hearing should understand at least 80% of spoken words in a typical broadband noise. As the S/N falls, intelligibility drops proportionately, so at 5 dB, it drops to about 70%, to 50% at 0 dB, and to 25% at −5 dB. People with NIHL may experience even larger reductions in intelligibility, while people accustomed to talking and communicating in noise do better.

Chances are, in an office environment, workers are collaborating much of the time and contributing to a shared output, product, or service. When noise interferes with voice communication, team-member efficiency is often impaired. After all, when this occurs, the time required to convey information is increased because more deliberate verbal exchanges become necessary. Not only is this annoying but it slows us down when fast action is needed and, even worse, can result in increased human error due to misunderstandings.

Clipping

When we communicate via telephone or other auditory transmission system, we also need to consider common forms of distortion. One of these is called "clipping" or "filtering," during which the peaks of the waveform are flattened off, cutting out certain frequencies or amplitudes. Here, the following findings apply: Frequency clipping affects vowels if the clipping occurs below 1000 Hz but has little effect if above 2000 Hz. Peak clipping is usually not critical if below 600 Hz or above 4000 Hz, because it primarily affects the vowels and reduces the quality of transmission only in a small and relatively benign manner. When applied to speech, it has surprisingly little effect on its intelligibility, with 80% or 90% of words still correctly interpreted by listeners even when the clipping is severe. On the other hand, center clipping is highly detrimental, particularly in the ranges of 1000–3000 Hz. Center clipping garbles the message, because it primarily affects consonants.

Improving Speech Communication

Speech communication is explicitly dependent on the frequencies and sound energies of any and all interfering noise. Several techniques exist to predict speech intelligibility based on narrowband noise measurements; examples include the Articulation Index, the speech interference level, or the simpler preferred speech interference level (please see the literature for more specific information). Using these measures, we can decide what kind of interfering sound must be eliminated and whether the speech requires improvement.

TABLE 9.2

International Spelling Alphabet

A: Alpha	N: November
B: Bravo	O: Oscar
C: Charlie	P: Papa
D: Delta	Q: Québec
E: Echo	R: Romeo
F: Foxtrot	S: Sierra
G: Golf	T: Tango
H: Hotel	U: Uniform
I: India	V: Victor
J: Juliet	W: Whiskey
K: Kilo	X: X-ray
L: Lima	Y: Yankee
M: Mike	Z: Zulu

Speech consists of five major components: the message itself, the speaker, the transmission, the environment, and the listener:

- The *message* becomes clearest if the context is expected, the wording is clear and to the point, and the ensuing actions are familiar to the listener.
- The *speaker* should use common and simple vocabulary with a limited number of terms; redundancy can be helpful as well. Talking speed should be slow, and if possible, it is best to use phonetically differentiated words. The International Spelling Alphabet is shown in Table 9.2.
- The *transmission device*, for example, the telephone, should be of "high fidelity," with little or no distortion in frequency, amplitude, or time.
- The *environment* or setting of the communications itself should be acoustically designed; specifically, make sure it has no noise or reverberation that would interfere with the message. There are two key guidelines related to the environment; the first concerns the environment of the transmission device, and the second has to do with the originating sound's environment:
  - First, in the transmitting device's environment, make sure to avoid or reduce generating any unwanted sound—make sure those around you are keeping their voices down, air flow is quiet, office equipment is quiet, and the office is efficiently laid out. "Active noise attenuation" is a promising new technique in which sounds become physically erased by instantaneously generated counter sounds that are of the same frequency and amplitude but of the opposite direction (180° off phase). This phase cancellation, or destructive interference (mentioned earlier), currently works best at frequencies below 1 kHz.
  - The second guideline concerns the environment around the originating message; you will want to impede the transmission of unwanted sound from the source to the listener. In occupational environments,

we can encapsulate the noise source, put sound-absorbing or sound-damming surfaces in the path of the sound, or physically increase the distance between noise source and ear. Another solution is to wear a hearing protection device (HPD) (plug, muff, helmet) that reduces the annoying subjective effects of sounds and avoids auditory harm. Of course, in general, if a condition exists in an office that necessitates wearing an HPD, we should first and foremost examine the environment and, if at all feasible, remedy the evidently hazardous condition.

- The listener's ability to *understand* the message is, of course, affected by existing noise—which should be eliminated. In unusual circumstances, it may be necessary to wear a special HPD that is penetrable by desired frequencies and to employ modern electronic devices that amplify certain bandwidths of sound depending on the existing amplitude and passively or actively suppress ambient noise. Also, the listener's hearing ability affects his or her facility in understanding a message. A number of devices exist to improve a person's hearing; you may want to reread the segment on hearing aids earlier in this chapter.

MUSIC WHILE WE WORK?

Music is probably one of the oldest art forms and earliest expression of human emotion. Music has long accompanied activities: consider workers singing during field labor or soldiers marching to rhythmic trumpets. The effects of music on industrial work were observed early in the twentieth century. Certainly, music can affect an individual's mood; even the most cynical among us would have difficulty arguing that a favorite song can elevate our spirits. It appears that music can help exercisers endure longer workouts or run an extra lap or two; some exercise routines thrive on loudly thumping soundtracks. But what about music's impact on productivity for those of us who work in an office?

Studies on the relationship between music and the brain have pointed to music's ability to help employees be more productive or help children learn more effectively. However, the type of music one listens to might not be as important as the listener's preference for it. Quite simply, individual preferences vary widely; some like to listen to "their" music while working, while others may find it distracting.

While music may seem like a tremendously difficult subject to study—after all, our preferences for music vary so widely—researchers can try to understand the link between music and how our body reacts to it. Consequently, although we still do not have much conclusive research on music's effect on office workers, we can make some general observations:

- For mundane, routine work, music can break up the monotony and improve output; whether this is because music accelerates performance or simply enhances our mood and thereby improves performance is unclear—of course, that may not matter if the outcome (higher productivity) is the same. Decades-old research showed that listening to music can make repetitive and boring tasks more enjoyable—and make workers tackling these tasks more productive. A 1995 study by researchers at the University of Illinois found that listening to music with headphones made workers more effective, and the largest gains in productivity were evident in simple, repetitive tasks. For example, in workers doing data entry, productivity jumped by 14% while lifting workers' moods (Portes, 2015).

- In a noisy workplace, where sounds are intrusive and distracting, music can be a respite because it can help mask the disruptive noise. Imagine being in an open workspace—increasingly common, as we explain in Chapter 4—where colleagues' conversations, loud sales calls, ad hoc meetings, and chirping cell phones are all taking place around you. Wearing headphones as you concentrate on your particular task can help drown out the environmental noise while allowing you to concentrate on your task.

- Certain music may be more suited to different kinds of work. Do you need to prepare yourself for a meeting during which you will be presenting your work? Perhaps opt for some fast-paced music or music written in a major key, which has shown to cause people to breathe faster, which is a physical sign of happiness (Edmonds, 2015). Do you need to relax after a stressful phone call? Consider slower music, written in a minor key, which seems to have a calming effect.

- Personal choice—once again—appears to be crucial when it comes to music at work. Consider that music helps some workers manage distractions and concentrate on their tasks, while for others, music feels overpowering. It seems that music tends to have a relaxing effect when people *choose* to listen to it but not when it is forced upon them. Importantly, music is personal. Our choice in music may irritate our colleague, and vice versa; the beauty of music is clearly "in the ears of the beholder."

- A *caveat*: Immersive lyrics—ones that compel us to listen and interpret the words—appear to be distracting when creativity and concentration are needed to complete the work or when we are attempting to learn a new or difficult concept.

ERGONOMIC DESIGN RECOMMENDATIONS

The following applies for office sound levels:

- The overall sound level in the office should be between 50 and 75 decibels (dB), ideally near 65 dB.
- For people doing work that requires intense concentration over long periods of time, the existing sound level should not change appreciably or dramatically; most people find changes invigorating but not if these changes are abrupt, startling, or otherwise unpleasant. We note here that what may be "abrupt, startling, or unpleasant" depends on the circumstances and is also highly individual.
- Some reverberation in the room is actually desirable—specifically, the reflection of sound at hard surfaces (floor, walls, ceiling, windows, furniture)—because it makes speech sound alive and natural.

If it is too loud for you, you should do the following:

- Eliminate the sound at its source:
  - Replace noisy equipment or machines with quieter ones.
  - Turn down the sound level of the ringer on telephones.
  - Ask your coworkers for quieter behavior.
- Reduce your exposure to the sound:
  - Move to a different office.
  - Place the noisy piece of equipment outside your room.
- Encapsulate the source of sound.

If there is too much reverberation in the office, "soften" hard surfaces that otherwise reflect the sound; use, for example, drapes, carpets, and acoustic tiles, which dampen or absorb sound.

If all else fails, do the following:

- Mask offending sounds by creating a "sound curtain" such as by playing music that is pleasant to you (perhaps through your personal earphones) or have a generator of "pink noise" (or "white noise") installed.
- Use personal hearing protectors (muffs or plugs).

If it is too quiet for you, do the following:

- Play music.
- Have other people move into your office.
- If your office is "sound dead" because there is too little reverberation, consider removing some drapery, carpets, acoustic tiles, and similar sound-deadening materials. Bare floors and walls, large pictures under glass, or windows have hard surfaces that reflect sounds.

Noise-Induced Hearing Loss (NIHL) may happen slowly, over time, or suddenly. Sudden NIHL results from exposure to short-duration sound of high intensity, like the sound an explosion produces.

Gradual NIHL results from exposure to everyday noises, and with continued sound exposure, more and more cilia are damaged, which the body cannot replace; also, nerve fibers that communicate from that region to the brain may degenerate, accompanied by corresponding impairment within the central nervous system.

Sound level, frequency, and duration of exposure (singly or repeatedly) are the critical descriptors for sounds that can damage hearing. In the United States, current OSHA regulations allow 16 hours of exposure to 85 dBA, 8 hours to 90 dBA, 4 hours to 95 dBA, etc. Other countries have different regulations and limits in terms of allowable exposure to noise. Ranges of noise, lengths of exposure, and resulting effects are shown in Table 9.1.

A number of hearing aids are presently available to improve hearing in those who have hearing loss not caused by nerve damage. It is important to note that a hearing aid will *not* restore normal hearing, so avoiding hearing loss is key. We should do what we can to minimize NIHL and protect our hearing as much as possible.

REFERENCES

Berger, E. H., Royster, L. H., Royster, J. D., Driscoll, D. P., and Layne, M. (2000). *The Noise Manual* (5th ed.). Fairfax, VA: American Industrial Hygiene Association.

Edmonds, M. (2015). Is there a link between music and happiness? http://science. howstuffworks.com/life/music-and-happiness.htm. Accessed December 5, 2015.

Kroemer, K. H. E. (2016). *Fitting the Human*. Boca Raton, FL: CRC.

Lin, F. (2011). Archives of internal medicine. One in five Americans has hearing loss. http://www.hopkinsmedicine.org/news/media/releases/one_in_five_americans_has_ hearing_loss. Accessed December 5, 2015.

National Institutes of Health (NIH). (2015). Hearing aids. *NIDCD*. http://www.nidcd.nih.gov/ health/hearing/pages/hearingaid.aspx. Accessed December 5, 2015.

Portes, S. (2015). Right music at work. *Bloomberg News*. http://www.bloomberg.com/news/ articles/2015-10-19/the-science-of-picking-the-right-music-at-work.

10 Office Climate

It doesn't matter what temperature the room is, it's always room temperature.

—Steven Wright, comedian

That feeling when you're so cold you'd give anything to be warm - I've had it before, literally huddled around a candle flame on an ice sheet.

—Bear Grylls, adventurer, writer, and television presenter

OVERVIEW

The body temperature of quite a few animals, such as fishes and salamanders, simply changes with their environment: when it is cold, they are cold, and they are warm when it is warm around them. These animals are somewhat misleadingly called "cold-blooded." We humans maintain a rather constant temperature of about 37°C at our core, especially in the brain and chest cavity of our bodies, regardless of the outside climate. Consequently, we call ourselves warm-blooded.

HEAT EXCHANGES

Keeping a constant core temperature is a complex task for the human body's temperature control system. The body itself generates heat energy and at the same time exchanges energy with the environment. When our surroundings are cold, we want to prevent excessive heat loss, and we can avoid losing heat by simply choosing warm clothing. When our environment is hot, our body must achieve an outward flow of heat to prevent overheating. This can be difficult to do, because heat always flows from the warmer to the colder. There are technical ways to help create that outward heat flow, with the most complete—albeit also most expensive—solution involving "climatizing" our environment: controlling its temperature, humidity, and air movement.

THERMODYNAMICS

Heat exchange between our environment and our body follows the thermodynamic processes of radiation, convection, conduction, and evaporation. The effectiveness of these processes depends on several conditions, especially our clothing's insulating properties and the energy level of the work we are performing at the time.

FEELING COMFORTABLE

For our well-being and comfort, the temperature difference between exposed skin and the environment is very important, but humidity and airflow also play major roles—as do our attitude and level of acclimatization. Within reasonable limits and given appropriate clothing, we humans can function well in both warm and cool environments.

Note: You may skip the following part and go directly to the "Ergonomic Design Recommendations" section at the end of this chapter—or you can get detailed background information by reading the following text.

WHAT IS A "CLIMATE"?

Some people use the term "office climate" when referring to how people in an office treat each other, work with each other, and get along with each other. These are very important psychological and sociological features, and we cover them in some detail in Chapter 2, but we bring them up here only to differentiate them explicitly from physical climate, which is what we address in this chapter.

Physical climate can be described in two ways: one is to actually scientifically measure the present air temperature, humidity, and air movement, which, together with energy generated and exchanged by radiation and conductance, objectively describe the environment's thermodynamics. The other aspect, and the more important one for the people in the office, is the "personal climate"—how each individual feels about these thermodynamic conditions: whether or not they are subjectively perceived as comfortable.

OFFICE CLIMATE AFFECTS US

Our body's temperature control system tries to keep us at a "neutral" temperature, so that we feel neither chilled nor too warm. Fortunately, in most office buildings, the climate conditions change very little, even as the outdoor environment shifts and changes, so the task of our body's thermoregulatory system is fairly simple. Understanding how the thermoregulatory system works helps engineers design appropriate climate control systems and allows us to select the most appropriate settings, clothing, and behavior to be physically comfortable with the climate in our office surroundings. Accordingly, we cover the basics of the human body's thermoregulatory system below.

OUR THERMOREGULATORY SYSTEM

Our body generates heat from the energy that we absorb from food and drink. This heat energy is circulated throughout the body by the blood. We exchange heat with our environment by gaining heat in a hot climate and losing it in the cold. That exchange takes place mostly through our skin and in our lungs. How much energy we exchange with our surroundings depends on the surface participating in the exchange, especially exposed skin. Clothing helps insulate the skin surface from heat transfer.

Whether or not we feel comfortable depends on many features; some are easy to measure, such as temperatures, while others are subjective and depend on how our body functions, such as our individual metabolism and circulation. Other factors that influence our comfort level include how we are dressed, how and how much we move around, and what season it is.

If our body's core temperature were to change even just a mere two degrees from its 37°C norm, our body's functions and resultant task performance would suffer tremendously. However, in a sheltered office environment, the human thermoregulatory system has no problem keeping the core close to 37°C. Although our core temperature remains relatively constant, at the skin, temperatures differ considerably from region to region. For example, our toes may be at 25°C, legs and upper arms at 31°C, and the forehead at 34°C all at the same point in time: for most of us, this feels comfortable.

Measuring Temperatures

Physicist Daniel Gabriel Fahrenheit (1686–1736) worked with thermometers and set the temperature of a mix of ice and water to 32°F and the temperature of boiling water 180°F higher, at 212°F. The eponymous Fahrenheit scale was useful for scientists, but in 1742, the astronomer Anders Celsius (1701–1744) suggested a metric scale with the freezing temperature of water set to zero and its boiling temperature to 100°C. This centigrade scale was called Celsius scale by international agreement in 1948 and is used in all countries except in the United States. For conversion of temperatures' degree values from one scale into the other, one must consider the different settings for freezing and boiling temperatures as well as the number of degrees between freezing and boiling of water (Figure 10.1):

Fahrenheit to Celsius: [°F − 32] [5/9] → °C

Celsius to Fahrenheit: [9/5°C] + 32 → °F

Of course, plenty of online calculators will quickly show conversions between Celsius and Fahrenheit.

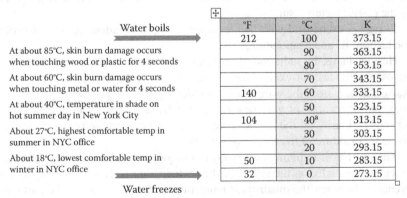

°F	°C	K
212	100	373.15
	90	363.15
	80	353.15
	70	343.15
140	60	333.15
	50	323.15
104	40[a]	313.15
	30	303.15
	20	293.15
50	10	283.15
32	0	273.15

Water boils

At about 85°C, skin burn damage occurs when touching wood or plastic for 4 seconds

At about 60°C, skin burn damage occurs when touching metal or water for 4 seconds

At about 40°C, temperature in shade on hot summer day in New York City

About 27°C, highest comfortable temp in summer in NYC office

About 18°C, lowest comfortable temp in winter in NYC office

Water freezes

[a]Core body temperature 37°C

FIGURE 10.1 Temperature scales in common usage. F, Fahrenheit; C, Celsius; K, Kelvin.

ENERGY BALANCE

Our body derives energy from the food and drink we ingest—and energy expenditure depends on how we move, how much we rest, how we interact with our physical environment, and how our body's cells and tissues perform. Energy expenditure is continuous, but it varies throughout the day. We can use a simple equation to describe the energy exchange between inputs to the body and outputs from the body:

$$I = M = H + W + S \qquad (10.1)$$

where I is the energy input via food and drink that the body transforms into metabolic energy M. The energy quantity M divides into the heat H that must be dispelled to the outside, the external work W done, and the energy storage S in the body.

Assuming for convenience that the quantities I, W, and S remain unchanged, we can concentrate on the heat energy exchange with the thermal environment according to

$$H = I - W - S \qquad (10.2)$$

The system is in balance with the environment if all metabolic energy H is dissipated to the environment. However, the case gets complicated by the fact that the body often receives additional heat energy from warm surroundings—or may lose heat in a cold environment.

TRANSMITTING ENERGY THROUGH RADIATION, CONVECTION, CONDUCTION, AND EVAPORATION

We do not live in isolation; instead, the human thermoregulatory system interacts with the environment. This is thermodynamically easy in a cold environment, but we may feel uncomfortably chilly; dissipating heat energy is more difficult in warm environs where we may feel overly warm. In a comfortable climate, which is what we find in most offices, the regulatory task is to dissipate the heat energy generated by the body's metabolism.

Energy is exchanged with the environment through radiation, R; convection, C; conduction, K; and evaporation, E.

Heat exchange by radiation, R, is a flow of electromagnetic energy between two opposing surfaces, for example, between a windowpane and a person's skin. Heat always radiates from the warmer to the colder surface. As a consequence, the human body can either lose or gain heat through radiation. The amount of radiated heat depends primarily on the temperature difference between the two surfaces but not on the temperature of the air between them.

The amount of radiating energy Q_R lost or gained by the body through radiation depends essentially on the size S of the participating body surface and on the difference Δ between the quadrupled temperatures T (in Kelvin) of the participating surfaces:

$$Q_R = f(S, \Delta T^4) \qquad (10.3)$$

This can be approximated by

$$Q_R \approx S \times h_R \Delta t \qquad (10.4)$$

where
 h_R is the heat-transfer coefficient
 t is the temperature in degrees Celsius

Figure 10.2 illustrates how heat from the sun warms our skin, without directly affecting the temperature of the air. Heat may also radiate from a fire, a hot windowpane, a heating radiator, or any other "warm body" whose surface temperature is higher than that of our skin. Of course, if the skin is warmer than, say, the cold pane of a window, and if there is no insulator such as clothing between the skin and the glass, then we radiate our heat to the window, as sketched in Figure 10.3. In this case, our body is losing heat, and we are growing colder.

 Heat exchange through convection, C, and conduction, K, both follow the same thermodynamic rule. The heat transferred is again proportional to the area of body surface participating in the process and to the temperature difference between our surface and the adjacent layer of the external medium. Hence, in general terms, heat exchange by convection or conduction is

$$Q_{C,K} = f(S, \Delta t) \qquad (10.5)$$

approximated by

$$Q_{C,K} \approx S \times k \times \Delta t \qquad (10.6)$$

with k as the coefficient of conduction or convection.

FIGURE 10.2 Heat radiated from the sun can feel so good, because it warms us. Of course, if we are already hot, we seek the shade.

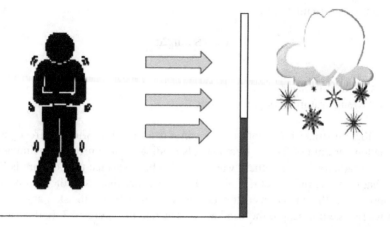

FIGURE 10.3   Our skin radiates heat to a colder surface, such as a cold windowpane in the winter. To avoid losing heat, we can cover up—either our body with clothing or the glass with a curtain, or both.

Heat exchange through convection, C, takes place when the body surface is in contact with air (or other gas or fluid). The body surface can be skin; it can also be the surface along the breathing pathways and in the lungs. Heat energy transfers *from* our body to the adjacent colder layer of gas (or fluid), or it transfers *to* our body if the surrounding medium is warmer.

Convective heat exchange is facilitated if air moves quickly along the body surface, which removes a stationary level of insulating air; this helps maintain a temperature and humidity differential. In the office, a fan can generate airflow.

Figure 10.4 shows how the air moving along our skin carries away the layer of warmed (and moist) air that is close to our skin. This makes convective heat exchange more effective, and if the air around us (or water, when we swim) is colder than the skin, we lose body heat. Think of how enticing it feels to jump into a cold pool for a refreshing dip on a hot summer's day or, conversely, how unappealing it feels to initially immerse yourself into a cool indoor pool in colder seasons when you are swimming laps for exercise. Keeping the air layer intact, such as when wearing loose-fitting clothes, acts as an insulator reducing the heat transfer. The same thermodynamic mechanisms are in effect when we are hot and sweaty and enjoy the cooling effect of air being blown at us from a fan or through a refreshing breeze, as sketched in Figure 10.5.

Conductive heat exchange, K, exists when the skin is in touch with a solid body. As long as there is a difference in temperature, heat naturally flows toward the colder body, quickly if the conductance of the piece is high, like in metal; less energy flows if the skin touches an insulator with a low value, like wood. Consider that wood on furniture feels "warm" because its heat-conduction coefficient is below that of human tissue. Metal of the same temperature accepts body heat easily and conducts it away; therefore, it feels colder than wood even if the temperatures of the wood and the metal are exactly the same.

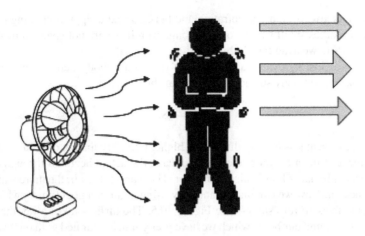

FIGURE 10.4   Airflow removes the layer of warmed air from our skin and makes us lose heat by convection to the colder fresh air. Stopping the current of air and using clothes that insulate (partly by keeping the warm layer of air trapped) prevent us from growing cold via convection.

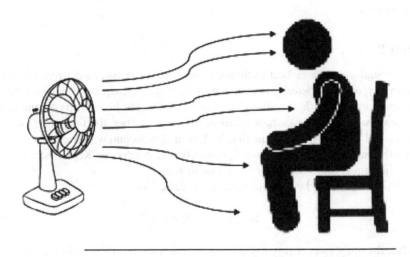

FIGURE 10.5   Airflow removes the layer of moist and warm air that our hot body created from our skin and makes us lose heat by evaporation of sweat, and by convection, to the drier and colder air blown in. This breeze is welcome on a hot day, not so welcome on a brisk one.

Heat exchange by evaporation, E, is in only one direction: the human loses heat by evaporation. There is never condensation of water on living skin, which would add heat. Evaporation of sweat, which is mostly water, requires energy of about 580 cal/cm³; this energy is primarily taken from the body and therefore reduces the heat content of the body by that amount. Some water is evaporated in the respiratory passages but most appears—as sweat—on the skin.

There is always some perspiration and hence sweat evaporation going on: that is why our clothes smell when worn too long, even if we are not engaged in strenuous activity while wearing the clothing.

The heat lost by evaporation Q_E from the human body depends mostly on the participating wet body surface, S, and on air humidity, h:

$$Q_E \approx f(S, h) \tag{10.7}$$

Evaporative heat loss is more difficult in higher humidity than in dryer air conditions. Movement of the air layer at the skin increases the actual heat loss through evaporation since it replaces humid air by drier air. The warmer the air, the more water vapor it can accept. Now we can understand why using a fan in a hot climate, which blows air at us, feels so refreshing—see Figure 10.5. The airflow takes away the layer of humid air around our body, which we have pretty much saturated with our evaporated sweat, and replaces it with dryer air. If the blown air is hot, we would overtly prefer a cool breeze, but warm air accepts our sweat more easily.

Because air's saturation with water depends on air temperature (besides atmospheric pressure), heat loss by evaporation depends indirectly also on the difference in temperature, Δt, as are the other heat transfers by radiation, convection, and conduction.

HEAT BALANCE

The actual amounts of heat exchanged with the surroundings depend, directly or indirectly, on the difference in temperature between participating body surfaces and the environment, as the equations listed earlier show. The body keeps its temperature constant when it achieves heat balance: this occurs when the metabolic energy H developed in the body (see Equation 10.2) is in equilibrium with the heat exchanged with the environment by radiation R, convection C, conduction K, and evaporation E. Equilibrium is achieved if they all add up to zero.

This condition of heat balance can be expressed as

$$H + R + C + K + E = 0 \tag{10.8}$$

When the body loses energy to the environment, one, several, or all of the quantities R, C, K, and E are negative, but they are positive if the body gains energy from the surroundings. E is always negative.

Temperature Regulation and Sensation

As mentioned earlier, if our body's core temperature changes about two degrees from its 37°C norm, our body's functions and its ability to perform tasks are severely and negatively affected; larger temperature deviations can kill us. Even in a sheltered office environment, where the human thermoregulatory system can easily keep the core close to 37°C, there are major temperature differences from one area of our body to another.

In the human body, heat is generated in "metabolically active" tissues—primarily by skeletal muscles, but also in internal organs, fat, bone, and connective and nerve tissue. The heat energy is circulated throughout the body by our blood. Blood vessel constriction, dilation, and shunting regulate the blood flow within the body. Heat exchange with the environment takes place at the body's respiratory surfaces (primarily in the lungs) but mostly through the skin. The equations shown earlier indicate that the energy exchange with our surroundings depends largely on the area of participating skin surface, especially skin that is naked and exposed to the environment. Clothing helps insulate the skin surface from heat transfer.

In a cold environment, body heat must be conserved. The body does this automatically by reducing blood flow to the skin—which keeps the skin colder—and we do it deliberately by increasing insulation via dressing in warm clothing. In a hot environment, body heat must be dissipated and heat gain from the environment must be minimized. The body does this by increasing the blood flow to the skin, making it warmer, and by increased sweat production and evaporation—and we do this deliberately by dressing more lightly. This may be surprising at first reading: a healthy body responds to a cold environment by making its skin surface colder and to a hot environment by making the skin warmer.

If the body is about to overheat, it needs to minimize internal heat generation. For this to happen, our bodies reduce muscular activities. Consider the warm-climate countries in which inhabitants routinely take a "siesta," a midday break. Such breaks evolved in large degree because people recognized their bodies' signals and rested their muscles accordingly. When their bodies began to overheat due to the peak heat of the day, taking a siesta would allow their bodies to sharply reduce muscular activity (through rest or sleep). In the opposite case, when more heat must be generated, increasing our work or exercise level and thus augmenting our muscular activities will help us warm up. If you have ever watched a foot race like a marathon held in cold weather, you would notice that the joggers dress in layers so that they can discard clothing as they warm up during their run. In fact, in highly competitive or elite races, many runners dress quite lightly and probably feel chilly at the beginning of their run, knowing that heat generation will warm them up quickly.

Hermann was taking his two small children skiing on this cold, bright Saturday, and the whole family was excited. He had carefully checked the weather forecast for the day and the hour-by-hour prediction called for a frosty beginning to the day, with a nice warm-up in temperature by midmorning. Hermann liked to get to the slopes early, right when they opened; he knew that his kids liked to get as many runs in as possible. He dressed them warmly in layers and with their warmest clothes, hats, gloves, and scarves. They set out early and were queuing up for the ski lift right at opening time. He looked down at his kids and saw that they were huddled together in their down parkas and ski pants, shivering, and he knew they were about to unleash a volley of whining. "Okay, boys," he said cheerfully, "let's see how many times you can jump up and down! Who can jump twenty times?" Giggling, the kids started hopping around, and their shivering soon stopped. Whew, he thought to himself. Whining was successfully averted!

Muscular, vascular, and sweat-production functions cooperatively regulate body heat in response to the surrounding climate. Our bodies have a number of temperature sensors, located in the core and the shell of the body. Heat sensors generate signals particularly in the range of approximately 38°C to 43°C. Cold sensors are most sensitive from about 35°C to 15°C. There is some overlap in the sensations of "cool" and "warm" in the intermediate range. Between about 15°C and 45°C, perception of either "cold" or "hot" condition is highly adaptable: under certain circumstances, we can grow accustomed to these temperatures. Below 15°C and above 45°C, human temperature sensors are less discriminating but also less adapting. A paradoxical effect occurs around 45°C, where sensors again signal "cold"—we may get goose bumps—while in fact the temperature is quite hot.

ACHIEVING THERMAL HOMEOSTASIS

In a sheltered environment, such as an office, the human body achieves thermal equilibrium (called homeostasis) primarily by regulating the blood flow from deep tissues and muscles to skin and lungs. In a less sheltered environment, our body also uses muscular activity to generate heat; it can do so involuntarily by shivering, or we can do so purposefully by exercising or working our muscles. Conversely, to avoid gaining heat, we reduce or abolish muscular activities.

CLOTHING, WORK, AND THERMAL COMFORT

Changes in clothing and in the climate of our shelter (office, home) represent our purposeful actions to establish thermal homeostasis. Together with blood-flow regulation and muscle activities, these efforts allow us to achieve the appropriate temperature gradient between the skin and the environment. How we dress and what our environment is like also affect radiation, convection, conduction, and evaporation. "Light" or "heavy" clothing differs in terms of permeability and ability to establish stationary insulating layers. Clothes also affect conductance, that is, energy transmitted per surface unit, time, and temperature gradient. Also, their color determines how much external radiation energy is absorbed or reflected. Especially in the sun, light-colored clothes reflect radiated heat, while dark surfaces absorb it. Consequently, on a hot and sunny summer's day, it may be a good idea to swap out that black baseball cap for a lighter-colored one.

Clothing, in and of itself, does not produce heat, of course, but it can prevent body heat from escaping into the environment. It does this by preserving the warm the layer of air that exists between skin and clothing. Air is a relatively bad conductor of heat, and therefore a good insulator. The insulating properties of clothing can be expressed in "clo" units, where one clo equals the thermal insulation required to keep a resting person indefinitely comfortable at a temperature of 21°C. This is approximately the value of the "normal" clothes worn by a sitting person at rest in an office where the air is at about 21°C and has about 50% relative humidity.

The "clo," a term derived from the word "clothes," is not a standard international unit (the standard international unit of thermal resistance is m² K/W, where 1 clo corresponds to 0.155 m² K/W (about the thermal insulation value R = 0.88), but it has the advantage of being easily understood: "one clo" is what a man wears when dressed in a shirt, trousers, suit jacket, and light underclothes; "2 clo" when in his ski clothes; "zero" when naked.

ASSESSING THE THERMAL ENVIRONMENT

In terms of physics and engineering, our thermal office environment is described by four factors:

1. Air temperature
2. Humidity
3. Air movement
4. Temperature of surfaces that exchange energy by radiation and conduction

The combination of these four factors determines the physical conditions of the climate and our perception of the climate.

Air temperature can be measured with thermometers, thermistors, or thermocouples. For exact measurements, we must ensure that the ambient temperature is not affected by the other three climate factors, particularly by humidity. To determine the so-called dry temperature of ambient air, we use a sensor that we keep dry, and we shield it with a surrounding bulb that reflects radiated energy. Consequently, we measure air temperature with a so-called dry-bulb thermometer.

Air humidity may be measured with a psychrometer, hygrometer, or other electronic devices. These devices usually rely on the fact that the cooling effect of evaporation is proportional to the humidity of the air; higher vapor pressure makes evaporative cooling less efficient. Therefore, we can assess humidity using two thermometers, one dry and one wet.

The highest absolute content of water vapor in the air is reached just before water droplets develop. The amount of possible vapor depends on pressure and temperature of the air: lower pressure and higher temperature allow more water vapor than lower temperatures and higher pressure. When we speak of humidity in percent, we are referring to "relative humidity," the actual percentage of vapor content in relation to the possible maximum content ("absolute humidity") at the given air temperature and air pressure.

Air movement is measured with various types of anemometers using mechanical or electrical principles. We can measure air movement with two thermometers as well—one dry and one wet (similar to what we would do to assess humidity), calibrated to the fact that the wet thermometer shows more evaporative cooling with higher air movement than the dry thermometer. Air moving at higher velocity cools human skin better (by convection and evaporation) than stagnant air.

Radiant heat exchange depends primarily on the difference in temperatures between the participating surfaces of the person and the surroundings, on the emission properties of the radiating surface, and on the absorption characteristics of the receiving surface. One practical way to assess the amount of energy acquired through radiation is to place a thermometer inside a black globe that absorbs practically all arriving radiated energy.

THERMAL COMFORT

While a person's feeling of comfort has much to do with the physics of thermal balance, discussed earlier, other factors influence our perception of physical comfort and well-being. These include skin dampness in a warm environment and skin temperature in a cold environment.

In the past, various techniques were used to assess the combined effects of some or all of the four environmental factors—remember, these are air temperature, humidity, air movement, and temperature of surfaces that exchange energy by radiation and conduction—and to express these in one model, chart, or index. They resulted in several empirical thermal indices, which are based on data compiled from the statements of subjects who were exposed to various climates. Most establish a "reference" or "effective" climate that "feels the same" as various combinations of the several climate components.

The so-called effective temperature is the combination of dry temperature, humidity, and air movement and applies to the office environment where people wear relatively light clothing and work at fairly low levels of physical effort. Such a climate index can be provided by instruments specially arranged to respond to climate components the way a human being would.

The wet-bulb globe temperature (WBGT) index is generated by an instrument with three sensors whose readings are automatically weighted and then combined. It weights the combined effects of four climate parameters as follows:
For outdoors, it is

$$WBGT = 0.7 \ WB + 0.2 \ GT + 0.1 \ DB \qquad (10.9)$$

where
  WB is the wet-bulb temperature of a sensor in a wet wick exposed to natural air current
  GT is the globe temperature at the center of a black sphere of 15 cm diameter
  DB is the dry-bulb temperature measured while shielded from radiation

For indoors, such as in the usual office or home, the equation is simpler:

$$WBGT = 0.7 \ WB + 0.3 \ GT \qquad (10.10)$$

The Bedford and the American Society of Heating, Refrigerating, and Air-Conditioning Engineers (ASHRAE) scales are widely used to assess thermal comfort; they yield similar results (Youle, 2005).

An interesting side note: If you are a woman who works in an office and you often feel cold, you are not alone. It turns out that most office buildings are kept at a temperature that is comfortable to most men. It has been that way for decades, because buildings all over the world adhere to an indoor temperature standard determined by a model developed in the 1960s by the ASHRAE. The model is based on factors like air temperature, airspeed, relative humidity, clothing, and the rate at which our bodies make heat (our metabolism). Therein lies the issue: in the 1960s, women were not considered as part of that standard. In general, women are smaller than men, and they tend to have less muscle and more fat—and muscle produces more heat than fat. Since women typically produce less body heat than men, they are more likely to feel chilled in the workplace. Also, women dress differently. Does this explain why many women prefer a warmer environment (of roughly 25°C) than men (about 22°C)? Women could benefit from setting the thermostat a bit higher in a warm climate: it could save some energy and men could dress a bit more lightly. Obviously, that higher temperature setting would cost more energy in a cold environment (Lewis and Kiersz, 2015).

REACTIONS TO HOT ENVIRONMENTS

Our body produces heat even in very warm environments and must dissipate it. Following the laws of physics concerning heat transfer, our skin temperature should be higher than that of our immediate environment, so that we can easily rid ourselves of our excess heat through convection, conduction, and radiation. If our body cannot transfer enough heat this way, our sweat glands are activated, and our skin is cooled as the sweat evaporates.

We all vary in the number of sweat glands we have and how they are recruited into action. The overall amount of sweat developed and evaporated depends very much on our clothing, environment, work requirements, and also each individual's acclimatization (discussed on page 215). Also, some of us just sweat more than others.

If our body still cannot transfer enough heat through blood distribution and sweat evaporation, it will reduce muscular activities to lower the amount of energy generated. This is the final and necessary action of the body when it is approaching a point where the core temperature would exceed a tolerable limit. (As mentioned earlier, taking a midday "siesta" during the hottest period of the day in a hot environment at noontime is a good example of a natural escape from overheating, especially when said siesta is taken in a cool spot.) If the body is forced to choose between unacceptable overheating and continuing to perform physical work, the choice will be in favor of core-temperature maintenance, which means reduction or cessation of work activities.

We note here that these reactions, forced upon our bodies by extreme environmental conditions, would be rare indeed in an office setting. So, we cover this topic just to complete the discussion of how our body reacts to hot environments.

SMELLS AND ODORANTS

Our senses of smell and taste are brought to us by our chemosensory system, and our nose is a complicated and fascinating part of that system.

Humans detect smells by inhaling air that contains odor molecules, which then bind to receptors inside the nose, relaying messages to the brain. Most scents are composed of many odorants; a whiff of coffee, for example, is made up of hundreds of different odor molecules. Once inside the nostrils, air molecules land on the olfactory epithelium, a mucus-covered tissue that lines the nasal cavity. The epithelium contains millions of olfactory receptors, neurons that are capable of binding with specific odor molecules to help identify certain smells. When you pour yourself a cup of coffee, an odor molecule floating up into your nose will find and subsequently bind to an olfactory receptor that is specifically designed to identify that molecule (HHMI, 2004). This message—oh, it's coffee!—is sent via electrical impulses to a microregion—known as glomerulus—in the olfactory bulb, which is then sent along to other parts of the brain.

For decades, the belief was that humans were capable of smelling around 10,000 different scents. This assumption has been turned onto its nose, so to speak, in the past decade. Recent research has shown that the nose appears to be capable of smelling up to one trillion scents, although the exact number is still under debate (Young, 2015). Odor molecules have a variety of different shapes that can fit into several receptors at once, which makes it possible for our nose to identify far more smells than the number of receptors it has (HHMI, 2004).

Smell has a number of purposes, including the following:

- *Taste*: Smell and taste together create flavor, which entices us to eat and enjoy food; while our tongues' taste buds can detect sweet, sour, salty, bitter, and umami, more complex flavors (like grapefruit or teriyaki) depend on smell.
- *Sensing danger*: Smell, like detecting the odor of fire, poisonous fumes, and spoiled food, allows us to avoid danger.
- *Memory*: Smell is known for triggering memories and emotions, likely because the glomeruli on the olfactory bulb are located in the brain's limbic system, which is often associated with memory and emotions; the olfactory bulb is also linked to the amygdala, which processes emotion, and the hippocampus, known for its role in learning (Gaines, 2015).
- *Social cues*: People emit chemicals in their bodily fluids, like sweat and tears, and these chemicals bring smells with them, which allow us to "read" their social cues.

Especially in a hot environment, smells can be a problem. Most women are better able than men to perceive odors (and, from the female perspective, this is a blessing *and* a curse). "Odor" is what we call the sensations resulting from odorants, which are chemical substances. Some of these substances are odorless, such as carbon monoxide. Most everyday odorants are mixtures of several or many basic chemical components, which generate complex odors to which different people, in different environments, and at different lengths of exposure react quite differently. Many odorants (often accompanying air contaminants) stem from the outgassing of building materials, from byproducts of running machines, from the industrial and traffic environment, and from human activity like smoking or consuming pungent food.

Not all odorants are external: the body also generates odors. Our skin has two main types of sweat glands: eccrine glands and apocrine glands. Eccrine glands, which can be found over most of the body, open directly onto the surface of the skin, and they are responsible for secreting sweat onto the surface of the skin, where it cools our body. This fluid is composed mostly of water and salt and is essentially odorless. Apocrine glands develop in areas abundant in hair follicles, such as arm-pits and groin, and they empty into the hair follicle just before it opens onto the skin surface. Body odor arises from the apocrine glands, which are small in infants but develop during puberty. Unlike eccrine glands, apocrine glands produce a milky fluid that most commonly is secreted when we are under emotional stress. This fluid is odorless until it combines with bacteria found normally on your skin. Skin bacteria break down the fats found in this milky secretion, and this can lead to a pungent odor (Mayo Clinic, 2014).

While we use our olfactory sense daily, describing a smell can be difficult, partly because a smell is not easily quantified. When you think about it, the sensation of a smell is described in relation to other smells, pleasant or not. You might describe the yeasty smell of a brewery "bread-like," for example, or liken a foul smell to the odor of rotten eggs. Assessing "odor annoyance" has been attempted in a manner similar to noise classification (Hangartner, 1987). One of the first quantitative studies was done in Danish offices: very little "smell pollution" came from people (who, in case you are wondering—we did—bathed on average 0.7 times a day and did not smoke) but far more emanated from smokers, materials in the office, and the ventilation system (Fanger, 1988).

Increased humidity increases odor intensity, and in some countries, we have guidelines for how much humidity—or vapor pressure—should be allowed in an office setting. To measure pressure, ergonomists use a method of millimeters of mercury—mm Hg—as a unit of pressure. 1 mm Hg is called 1 torr, so 1 torr his-torically was the amount of pressure that literally produces a 1 mm difference in the heights of two columns of mercury in a manometer. In North America, vapor pressure in the office air should be between about 10 and 15 torr, and about 0.5 m^3 (17 ft^3) of outside ("fresh") air should be provided per minute for each person (this is according to AINSI/ASHRAE Standard 62.1-2004—please check your current local standards).

The effects of odors can be physiological, independent of the actual perception. When strong, they may stimulate the central nervous system, eliciting changes in body temperature, appetite, and arousal. Odor sensitivity and response to odors dif-fer from one person to the next, and for those people who are more sensitive to odors, detecting a foul odor can result in headaches and nausea. People who tend to be more sensitive to odors include pregnant women and people with chronic health issues, as well as people who have chemical hypersensitivities.

The most common health complaints resulting from exposure to odor-producing chemicals are

- *Upper respiratory*: scratchy throat
- *Lower respiratory*: coughing, wheezing
- *Eye irritation*: watery eyes, scratchy feeling
- *Gastrointestinal*: nausea, vomiting, diarrhea
- *Central nervous system*: drowsiness, dizziness, headaches
- *Cardiovascular*: tachycardia (increased heart rate), increased blood pressure
- *Psychological*: mood changes or fluctuations, behavioral changes

Odors also can have psychological and psychogenic effects that concern especially attitude and mood, including anger or benevolence toward others, cooperation, creativity, self-perception, and performance. They may be part of the "sick building syndrome" (Ballard, 1995; Fanger, 1988).

Sick building syndrome is a term that was first used in 1986 to describe how the compounds in the building (cigarette smoke, chemical outgassing from structural components, or biologics such as bacteria and fungi) interact with the building inhabitants in a way that causes some of them to develop symptoms (Davis, 2015). Also known as environmental illness or multiple chemical sensitivity, the concept of sick building syndrome is controversial.

Those who believe that sick building disease exists explain that causes of the syndrome are varied and multiple, and reaction to these causes depend on the individual—does he or she have a preexisting medical condition such as asthma? Office workers who feel they suffer from the syndrome may be more sensitive to low concentrations of some compounds and may have heightened immune response to such compounds.

Others who say there is no evidence for this syndrome agree that certain chemicals, biologics, and physical agents found in some buildings can cause disease, but once these are identified (for example, lead or asbestos), then the disease is identified and is not a new "syndrome" (Davis, 2015).

The U.S. Environmental Protection Agency (EPA) uses the term "building-related illness" as a term to describe known causes of problems—issues including toxic gasses and molds, mildew, bacteria, plants, and other known compounds found in buildings. According to the EPA, World Health Organization statistics indicate that as many as 30% of all buildings worldwide, new or refurbished, have air quality problems. The existence of such indoor air quality problems can cause complaints, negative health effects, and even result in lawsuits. However, these problems are identifiable and most can be remedied by such methods as using HEPA filters to reduce or eliminate most airborne particles, avoiding building air intakes located near sources of vehicle exhaust fumes or other irritants, and avoiding fungal and bacterial contamination of air-conditioning or other air circulating methods. According to the EPA, some people who are termed as having sick building syndrome are actually in a situation where they have a building-related illness that has not been investigated or the source identified.

REACTIONS OF THE BODY TO COLD ENVIRONMENTS

The human body has few natural defenses against a cold environment. Most of our counteractions are behavioral, such as putting on heavy and insulating clothing or using external sources of warmth. In a cold climate, the body must conserve heat. To diminish the flow of energy to the surroundings, the body naturally lowers the temperature of the skin, which reduces the temperature difference against the outside. Keeping circulating blood closer to the core, away from the skin, accomplishes this; for example, the blood flow in the fingers may be reduced to 1% of what it would be in a moderate climate. This explains why, in cold environments, our fingers and toes grow uncomfortably cold most quickly if we have not buffered them with warm gloves and socks. Developing "goose bumps" on the skin is another involuntary bodily reaction to cold; this physiological phenomenon is more helpful to fur-bearing animals to help retain a relatively warm layer of stationary air close to the skin. Given our relative dearth of body hair, the clothing we wear creates that sort of a stationary envelope: it acts like an insulator, reducing energy loss at the skin.

The other major reaction of the body to a cold environment is to increase metabolic heat generation. We can do this through purposeful muscular activities, such as moving segments of our body—think of a pianist flexing the fingers or runners at the starting line of a race jumping up and down. Such dynamic muscular work may easily increase the metabolic rate to 10 or more times the resting rate.

Metabolic heat generation often occurs involuntarily through shivering. The onset of shivering is normally preceded by an increase in overall muscle tone in response to body cooling. Right before we shiver, we usually experience a general feeling of stiffness. Then suddenly shivering begins, caused by muscle units firing at different frequencies of repetition and out of phase with each other. Since no mechanical work is done on the outside, the total activity is transformed into heat production, allowing an increase in the metabolic rate of up to four times the resting rate.

Do drafts make us sick? Some people believe they do: for example, when in Germany somebody shouts "es zieht!" (a draft!), then the offending window is shut, even on a hot day; no questions asked. Perhaps your grandmother advised you to steer clear of drafts in order to avoid getting a cold. But can a draft really result in dire health consequences? Not at all. Movement of clean air does not make you sick, and even getting wet and cold does not make you "get a cold." That said, contaminated air moving around you absolutely can cause health problems: somebody with a cold, or the flu, breathing and sneezing on you can indeed infect you.

Drafts can, of course, make us uncomfortable, and that brings us to the topic of windchills. Windchill is an important consideration for those who will be spending time outside in cold weather because a simple degree measure does not describe how cold we may feel. Imagine being outside in a blizzard; you will be far colder in bracing, icy winds at a temperature of −4°C than you would be on a still, breeze-free day at that same temperature. A suitable index of the "windchill" will indicate not only how we feel outside but also how long we can tolerate the cold. One of the first windchill measures, called Siple–Passel (named after the two explorers who derived it), was based on observing how quickly bottles of water froze in the Antarctic wind. Later indices attempted to apply windchill factors to human physiology; Figure 10.6 shows a commonly used windchill chart.

For combinations of temperature and wind speed, the windchill temperatures in Figure 10.6 show how an individual would feel. In a very light wind, for example, at a temperature of −15°C with a wind speed of 10 km/h, it would feel as if it were −21°C. If the wind would pick up to 50 km/h, the perceived temperature would fall to −29°C. We note that these windchill indices do not take into account the warming properties of the sun; bright sunshine could make it feel considerably warmer than the windchill index would indicate.

Actual air temperature (Celsius)

Wind speed (km/h)	5	0	-5	-10	-15	-20	-25	-30	-35	-40	-45	-50
5	4	-2	-7	-13	-19	-24	-30	-36	-41	-47	-53	-58
10	3	-3	-9	-15	-21	-27	-33	-39	-45	-51	-57	-63
15	2	-4	-11	-17	-23	-29	-35	-41	-48	-54	-60	-66
20	1	-5	-12	-18	-24	-30	-37	-43	-49	-56	-62	-68
25	1	-6	-12	-19	-25	-32	-38	-44	-51	-57	-64	-70
30	0	-6	-13	-20	-26	-33	-39	-46	-52	-59	-65	-72
35	0	-7	-14	-20	-27	-33	-40	-47	-53	-60	-66	-73
40	-1	-7	-14	-21	-27	-34	-41	-48	-54	-61	-68	-74
45	-1	-8	-15	-21	-28	-35	-42	-48	-55	-62	-69	-75
50	-1	-8	-15	-22	-29	-35	-42	-49	-56	-63	-69	-76
55	-2	-8	-15	-22	-29	-36	-43	-50	-57	-63	-70	-77
60	-2	-9	-16	-23	30	-36	-43	-50	-57	-64	-71	-78
65	-2	-9	-16	-23	-30	-37	-44	-51	-58	-65	-72	-79
70	-2	-9	-16	-23	-30	-37	-44	-51	-58	-65	-72	-80
75	-3	-10	-17	-24	-31	-38	-45	-52	-59	-66	-73	-80
80	-3	-10	-17	-24	-31	-38	-45	-52	-60	-67	-74	-81

FIGURE 10.6   Windchill chart, Celsius.

Expressed as a formula, for air temperatures lower than 10°C and wind speeds greater than 4.8 km/h

$$Windchill = 13.12 + 0.6215T - 11.37(V^{0.16}) + 0.3965T(V^{0.16})$$

where
     T is the air temperature in °C
     V is the wind speed in km/h

When the air around us is colder than our skin's temperature, we are losing heat through convection to the air from any skin surfaces that are not covered by insulating clothing. However, if we step outside and a breeze blows over our bare face, the heat loss increases because the wind removes the warm layer of air that surrounds our skin; this layer of warm air acted almost like a scarf. (If our skin is wet, there is additional cooling due to evaporation of that moisture.) The stronger the wind, the faster the cooling; what is more, the lower the temperature, the more impact the wind has.

Importantly, while high winds can make us feel chilled far more quickly, these windchill effects apply only to living creatures—they do not have the same impact on inanimate objects. More wind *does* mean that inanimate objects will cool to the air temperature more quickly—this is how the explorers Siple and Passel were able to perform their water bottle experiments to discover windchills—but even the highest winds cannot force an object's temperature below the air temperature. For example, if you are standing outside at 4°C, clutching a filled water bottle, and the wind is gusting by, chilling your exposed skin to feeling like it is –2°C, your water bottle will still be a relatively balmy 4°C and will not freeze.

Jose, a recent college graduate from Southern Florida, has just relocated to Chicago. On a blustery day in February, just a few days after moving to the Midwest, Jose has an interview with a CPA firm downtown. As he dresses that morning, he consults his smartphone to get the forecast; it informs him that temperatures are around 35°F (1.6°C). He faces a 30-minute commute on public transportation, and he wants to make sure he selects the most appropriate clothing. After all, since he is most accustomed to the sunny and warm climes of Southern Florida, he wants to avoid freezing as he makes his way downtown. He throws on an overcoat but forgoes hat, scarf, and gloves, thinking that the coat will suffice. By the time he walks to the elevated train stop and stands on the platform for 10 minutes, he is shivering and shaking from the cold. The whistling wind is howling around him as he blows on his freezing fingers and huddles closer to the heating lamps on the platform. The woman next to him watches his gyrations, amused, and takes in his lightweight overcoat. Jose notices her perusal. "I made a point of checking the forecast this morning," he grumbles, "and it's supposed to be 35°F. This feels a lot colder than that." She nods. "Well, sure, it is a lot colder than that. It's got to be no more than 25°F when you factor in the windchill." She sees the blank look on Jose's face. "Come on, now. Haven't you ever heard of windchill factors?" the woman asks with a chuckle.

ACCLIMATIZATION

Continuous or repeated exposure to hot and (to a lesser degree) cold conditions brings about a gradual adjustment of body functions, resulting in a better tolerance of the climate stress. This process is called acclimation, the result acclimatization.

Acclimation to heat increases sweat production, lowers skin and core temperatures, and reduces heart rate, compared with a person's first reactions to heat exposure. The process shows pronounced results in about a week, and full acclimatization is achieved within about two weeks. Interruption of heat exposure by just a few days reduces the effects of acclimatization; upon return to a moderate climate, acclimatization is entirely lost after about two weeks.

A healthy person can adjust to some heat, both dry and humid. Heat acclimatization does not depend on the type of work performed, whether heavy and short or moderate but continuous. A person who is healthy and well trained acclimates more easily than a person who is in poor physical condition, but training cannot replace acclimatization. Additionally, our bodies can adapt to heat, but not to dehydration, so it is important to drink enough water, both during acclimation and then throughout heat exposure, to replace fluid lost through sweat evaporation (Carroll, 2015).

Ester works as a marketing executive at an international textiles manufacturing company headquartered in Quebec. It has been a long, cold winter, and now, in March, it still seems like the chilly, snowy weather will last for at least several more weeks. For once, though, Ester does not mind the lingering winter; she has just arrived in Mexico for a 7-day honeymoon with her new spouse. After a leisurely brunch at their luxurious hotel on their first full day, they decide to take a long walk into town and back. Within 20 minutes of their stroll in the midday sun, however, Ester is feeling quite winded and very warm, and she calls out to her spouse, who is walking ahead of her, to slow down. Just then, she notices that they are passing by a crew of laborers working to lay the foundation of a new hotel close to town. She notes with some amazement that the men on the crew are working seemingly tirelessly, effortlessly carrying and moving large pieces of building material around as though those blocks of concrete were made of foam rather than stone and asphalt. Ester catches her breath as they watch and wonders out loud how the men are capable of functioning like that in what seems to her as oppressive heat. "I guess," Ester surmises, "they're just used to it."

Acclimatization to cold is much less pronounced than acclimatization to heat; in fact, there is some doubt that true physiological adjustment to moderate cold actually takes place when appropriate clothing is worn. In other words, since we dress for cold weather by wearing protective clothing and accessories like hats and gloves, our body may not have need to acclimatize.

As in a hot climate, the body must maintain its core temperature near 37°C in a cold environment. When exposed to cold, the human body first responds by peripheral vasoconstriction (constriction of blood vessels), which lowers skin temperature, in order to decrease heat loss through the skin. This can lead to local acclimatization in terms of the flow of blood in the hands and face. However, the blood supply to the head (brain) remains intact, so there is a risk that substantial heat loss will take place there unless warm hat and face mask provide cover. The predominant adjustment to cold conditions is in choosing proper clothing and engaging in physical work with the result that, in "normally cold" temperatures, it is not necessary that the body change its rate of heat production or, relatedly, of food intake. In other words, acclimatization to cold temperatures appears to be the result primarily of people becoming "smarter" at dealing with it—by learning to dress appropriately, for example, and engaging in physical movement.

There are no great differences between females and males with respect to their ability to adapt to either hot or cold climates, with women possibly at a slightly higher risk for heat exhaustion and collapse and for cold injuries to extremities. However, these slight statistical tendencies can be easily counteracted by ergonomic means and may not be obvious at all when observing only a few individuals of either gender.

EFFECTS OF HEAT OR COLD ON MENTAL PERFORMANCE

It is difficult to evaluate the effects of heat (or cold) on mental or intellectual performance, because there are few practical yet objective testing methods. Moreover, the range of subjective variations among individuals is huge, even as the climate fluctuations in offices are minor. However, as a rule, mental performance deteriorates with rising room temperatures, starting at about 25°C for the nonacclimatized person. That threshold increases to 30°C or even 35°C if the individual has acclimatized to heat. Brain functions are particularly vulnerable to heat; even if body temperature must be increased, keeping the head cool improves tolerance. A high level of motivation may also counteract some of the detrimental effects of heat. Thus, in laboratory tests, mental performance is usually not significantly affected by heat as high as 40°C WBGT (Kroemer, 2010; Kroemer et al., 2001).

Extreme cold conditions are not often found in offices. In the unlikely case of body core temperature dropping below the mid-30°C, vigilance and mental performance are reduced and nervous coordination suffers. Manual dexterity is reduced if finger skin temperatures fall below 20°C. Tactile sensitivity is reduced at about 8°C, and near 5°C, skin receptors for pressure and touch cease to function; the skin feels numb at these low temperatures—think of writing down a note on paper with seemingly immobile fingers.

CLIMATE CONTROL IN OFFICES AND OFFICE BUILDINGS

Throughout the 1800s, the architecture of large office buildings, and the arrangement of the offices within, still followed the example of the Uffizi ("offices") in Florence, Italy. Between 1560 and 1581, they were built in U form around an open court so that every room had a window to the outside, to provide natural lighting and ventilation.

One of the authors (Karl K.) remembers, with a bit of longing and nostalgia, the immigration office in Bombay (now Mumbai), India, in 1983: it was a long, low-slung, two-story building with wide verandahs in a parklike setting, all windows open to the breezes, and ceiling fans turning continuously—which made paperweights necessary to keep the stacks of document in place.

In such old large buildings, heating was done, if needed (not in Mumbai!), by coal fires and stoves in the rooms, and bad odors abounded when windows had to be kept closed. Design plans in forms of U, O, E, I, H, and T were employed into the 1930s to allow windows to the outside in every room (Arnold, 1999).

In 1906, Frank Lloyd Wright's Larkin Administration Building was constructed with "sealed ventilation" for some cooling and heating of the air. The Milam Building in San Antonio, Texas, was the first to be fully air-conditioned in 1936. This technology allowed for huge, city-block wide edifices, with many interior rooms located away from outside walls. The one edifice that really utilized all the new technology was Frank L. Wright's Johnson's Wax Administration Building, completed in 1939. It was entirely sealed and fully air-conditioned, and featured below-floor heating and clerestory windows constructed from bundles of glass tubes that provided diffuse light to rooms.

Air-conditioning was increasingly used in large office buildings, which, coupled with electric lighting, led to large and windowless buildings. The architectural and engineering evolution was interrupted by the Second World War, after which it resumed at full speed. Air-conditioning of large and small office buildings became commonplace in North America. In the course of this development, many concerns about the office climate grew as well: about the spread of tobacco smoke and other air contaminants, even diseases, by forced airflow, about "acceptable" indoor air quality, and about the work performance in offices that are too cool or too warm.

Please note that we cover the history of office buildings in more detail in Appendix B.

ERGONOMIC DESIGN RECOMMENDATIONS

There are many ways to generate a thermal environment that meets the physiological needs of people as well as their individual preferences. The primary approach is to adjust the physical conditions of the climate (temperatures, humidity, and air movement), which in turn influence the heating or cooling of the body via radiation, convection, conduction, and evaporation. The body's interactions are complex and must be carefully considered when designing and controlling the environment—guidance may be taken from ASHRAE recommendations in the United States, from international standards such as those outlined by ISO, and from national regulations, regional recommendations, and customs.

What is of importance to the individual is not the climate in general, the so-called macroclimate, but the climatic conditions with which one interacts directly. Every person desires an individual microclimate that feels "comfortable" under given conditions of adaptation, clothing, work, and individual preference.

The "suitable" microclimate is not only highly individual—we all define "suitable" on our own terms—but also variable. It depends, for example, on age (older people tend to be less active, to have weaker muscles, to have a reduced caloric intake, and to start sweating at higher skin temperatures), on the surface-to-volume ratio (larger in children than adults), and on the fat-to-lean body-mass ratio (generally larger in women than in men).

Thermal comfort depends also on the type and intensity of work performed. Physical work in the cold leads to increased internal heat production and hence to decreased sensitivity to the cold environment, while intensive physical work in a hot climate can become intolerable if an energy balance cannot be achieved. Similarly, sedentary work in a too-cool environment can make us feel chilly far more quickly than if we are moving around and thereby generating internal heat.

The clothing that we choose to wear also affects the microclimate; our clothing selection determines the surface area of exposed skin. More exposed surface areas allow for better dissipation of heat in a hot environment but can lead to excessive cooling in the cold. Air bubbles contained in the clothing material or between clothing layers provide insulation, both against hot and cold environments. This is why wearing appropriate layered clothing can help us stay warm in the winter. Permeability to fluid (sweat) and air plays a role in heat and cold. Even the colors of the clothing we choose play a role in regulating our temperature in a heat-radiating environment, such as in sunshine, where we will feel more comfortable in light-colored clothes. Darker colors absorb radiated heat while lighter colors reflect incident energy.

Convection heat loss increases if air moves swiftly along exposed surfaces. Therefore, with increased air velocity, body cooling becomes more pronounced. On a warm day, you will likely feel more comfortable if a breeze exists rather than if the air is stagnant, even if the temperature itself is unchanged. Conversely, on a very cold day, windchills can make us feel significantly more miserable than if the air were still.

When the air around us is colder than our skin's temperature, we are losing heat through convection to the air from any skin surfaces that are not covered by insulating clothing. This heat loss increases in wind because the wind removes the warm layer of air that surrounds our skin. The stronger the wind, the faster the cooling; what is more, the lower the temperature, the more impact the wind has.

Thermal comfort is also affected by acclimatization, a state in which the body (and mind) have adjusted to changed environmental conditions. A climate that felt rather uncomfortable during the first day of exposure may be much more agreeable after a couple of weeks. Seasonal changes in climate, unusual work, different clothing, and an evolving attitude all have major effects on what we are willing to accept or even to consider comfortable. In the summer, most people find warm, breezy, and even somewhat humid conditions comfortable, while during the winter we expect cool and dry weather: this is what many of us are accustomed to.

For these reasons, various and variable combinations of the climate factors—temperature, humidity, and air movement—can subjectively appear very similar. The wet-bulb globe temperature (WBGT) index takes into account temperature, humidity, wind speed, sun angle; it is most often used to assess the effects of warm or hot climates on the human. For a cold climate, various similar approaches have been proposed, but are not yet universally accepted (Youle, 2005).

Our sense of smell is invaluable to us in our ability to taste, sense danger, stimulate memory, and offer social cues. Because our sense of smell is so powerful, it also means odors in the office can be problematic. Unpleasant smells in the office are subjective—odors one individual tolerates or even enjoys can elicit a very different reaction from another person. For those people who are more sensitive to odors, detecting an unpleasant smell can have physiological and psychological consequences. Odors are particularly problematic for those with respiratory diseases or other chronic health issues.

ERGONOMIC RECOMMENDATIONS

With appropriate clothing and light physical work in the office, comfortable environment temperature ranges are about 21°C to 27°C in a warm climate or during the summer, but lower—between 18°C and the middle 20s (18°C–24°C)—in a cool climate or during the winter.

- The difference between air temperatures at floor and head levels should be less than about 6°C.
- Differences in temperatures between body surfaces and surfaces on the side (such as walls or windows) should not exceed approximately 10°C.

The preferred range of relative humidity is from 30% to 60%, ideally 40% to 50%.

Air velocity should not exceed 0.5 m/s and preferably remain below 0.1 m/s. Airflow should not generate sound levels above 60 dB.

Deviations from these zones are uncomfortable, can make work difficult, and may even become intolerable.

If the sun shines onto employees, particularly on warm days, they should be able to move out of the sun or to get into the shadow of blinds, curtains, screens, and the like.

WHAT TO DO IF YOU ARE NOT COMFORTABLE

If you feel too warm

- Lower the room temperature.
- Move away from a heat source such as a radiator and a warm wall or window; move out of direct or indirect sun exposure.
- Move closer to a cool surface.
- Lower air humidity (use a dehumidifier).
- Increase the air movement around you (unless the air is very hot).
- Wet your exposed skin; place a cool/humid piece of cloth on your forehead, neck, or wrist.
- Take off a layer of clothing; bare more skin.
- Keep your body at rest; do not exercise it.

If you feel too cool

- Increase the room temperature.
- Move closer to a heat source such as a radiator and a warm wall or window; move into the sunshine.
- Move away from a cool surface.
- Move closer to a warm surface.
- Decrease the air movement around you (unless the air is nicely warm).
- Add a layer of clothing; cover more skin.
- Keep your body moving.

If you feel too dry (dry throat, nose)

- Increase air humidity by evaporating water on a warm surface, using a humidifier; also drink water.

REFERENCES

Ballard, B. (1995). How odor affects performance: A review. In *Proceedings, ErgoCon'95, Silicon Valley Ergonomics Conference and Exposition.* San Jose, CA: San Jose State University.

Carroll, A. E. (2015). No, You Do Not Have to Drink 8 Glasses of Water a Day, In *New York Times 24 Aug,* http://www.nytimes.com/2015/08/25/upshot/no-you-do-not-have-to-drink-8-glasses-of-water-a-day.html?_r=1

Davis, C. P. (2015). Sick building syndrome. MedicineNet.com. http://www.medicinenet.com/sick_building_syndrome/article.htm. Accessed November 15, 2015.

Fanger, P. O. (1988). Hidden olfs in sick buildings. *ASHRAE Journal* 30(11), 40–43.

Gaines, J. (2015). Smells ring bells: How smell triggers memories and emotions. *Psychology Today*. https://www.psychologytoday.com/blog/brain-babble/201501/smells-ring-bells-how-smell-triggers-memories-and-emotions. Accessed November 15, 2015.

Hangartner, M. (1987). Standardization in olfactometry with respect to odor pollution control; Assessment of odor annoyance in the community. In *80th Annual Meeting of the APCA*, New York.

HHMI. (2004). Richard Axel and Linda Buck awarded 2004 Nobel prize in physiology or medicine. Howard Hughes Medical Institute. http://www.hhmi.org/news/richard-axel-and-linda-buck-awarded-2004-nobel-prize-physiology-or-medicine. Accessed November 15, 2015.

Lewis, T. and Kiersz, A. (2015). Why women are always freezing in offices, According to science. *Business Insider*. http://www.businessinsider.com/why-women-are-always-freezing-in-office-buildings-according-to-science-2015-7. Accessed November 15, 2015.

Youle, A. (2005). The thermal environment. In K. Gardiner and J. M. Harrington (Eds.). *Occupational Hygiene* (3rd ed.). Oxford, U.K.: Blackwell.

Appendix A: A Brief History of Ergonomics

From the earliest time on, humans selected objects found in the environment and discovered ways to use them as tools or weapons—fist-sized hunks of rock were perfect for pounding and long sharp pieces of wood made workable spears. They then began shaping or altering those objects on purpose so they would work better and fit both the human and the task at hand more appropriately—scoops made out of antelope bones, or sharpened rock topping a spear. Subsequent humans figured out how to create products from raw materials and manufacture them to meet a specific need or perform a given task. We can point to the Industrial Revolution, when machines like spinning jennies, which produced yarn, and rolling mills, which flattened iron ore into sheets, were developed to improve work flow and sharply increase output. Fundamentally, constructing items, building shelters and homes, and manufacturing and tailoring clothing are all true "human engineering" activities.

As human society grows more complex, organizational and managerial challenges develop, and human factors projects become all the more comprehensive. Consider the technique of "drying the legions," which the Roman Empire employed, whether consciously or through trial and error. Based on the principle of training and adapting the physiological capabilities of the recruits to the physical requirements of warfare, when the recruits no longer exhibited sweat on their skin—when they were "dry"—they were fit. We now have sophisticated and high-tech ways to measure fitness, but the basic concept of observing humans and designing systems around them is fundamentally the same.

Building the pyramids of ancient Egypt, assembling and training armies for warfare, sheltering the inhabitants of cities, and supplying them with food and water are all major projects that require careful planning and complex logistics together with sophisticated knowledge of human needs and desires.

EVOLUTION OF DISCIPLINES

Medical experts and artists have helped forge the way as pertains to human factors—they have always been most interested in physical build and performance, with a focus on anatomy, physiology, and anthropology. Consider that around 400 BC, Hippocrates, the "father of medicine," described a scheme of four body types—"humors"—that formed a person's character and health profile. Under this theory, people required an even balance of the four body fluids: blood, phlegm, yellow bile, and black bile. Hippocrates, Plato, and Aristotle had differing philosophies, but all three saw health as an equilibrium of the body as determined by humors.

Over the following centuries, more exact information accumulated into specialized disciplines. In the fifteenth to seventeenth centuries, highly gifted individuals like Leonardo da Vinci and Alfonso Giovanni Borelli seemingly mastered all the existing knowledge of anatomy and physiology and generated related engineering designs ("biomechanics"): they were artists, scientists, and engineers in one. Giovanni Alfonso Borelli is often described as the father of biomechanics. His De Motu Animalium, published in 1680, extended to biology the rigorous analytical methods developed by Galileo in the field of mechanics; he contributed to the modern principle of scientific investigation by continuing Galileo's custom of testing hypotheses against observation. Borelli was the first to understand that the levers of the musculoskeletal system magnify motion rather than force, so that muscles can produce much larger forces than those resisting the motion. Da Vinci's natural genius crossed so many disciplines that he epitomized the term "Renaissance man." Da Vinci studied nature, mechanics, anatomy, physics, architecture, weaponry, and more, often creating accurate, workable designs for machines like the bicycle, helicopter, and an airplane based on observations including the flying capabilities of a bat.

In the eighteenth century, the sciences of anatomy and physiology diversified and psychology began to develop as a separate discipline. For many decades, these sciences tended to be oriented toward theory, "pure" sciences, with little regard to practical applications; clinicians worked simply to understand the vagaries and complexities of the human being. The Industrial Revolution and the associated mass employment of workers, together with the already existing interest in deploying humans for military needs, ushered in "applied" aspects of the formerly pure sciences.

In France, during the early 1800s, Lavoisier, Duchenne, Amar, and Dunod researched energy capabilities of the working human body. Marey developed methods to describe human motions at work, and Bedaux studied work payment systems. The word "ergonomics" was first used by Wojciech Jastrzebowski in a Polish newspaper in 1857 (Jastrzebowski, 1857). The word comes from the Greek *ergo* (work) and *nomos* (rules, law). At the beginning of the twentieth century, Frank and Lillian Gilbreth developed the concept of time-and-motion study and divided human movement into small microelements they called "therbligs." In 1903, Frederick Taylor defined the scientific study of work in his publication "Shop Management"; he proposed enhancing productivity by making workers' movement patterns as simple as possible. The resulting discipline was known as Taylorism.

In England, the Industrial Fatigue and Research Board considered theoretical and practical aspects of human work. In Italy, Mosso constructed dynamometers and ergometers in his quest to understand fatigue. In Russia, Pavlov studied circulation, digestion, and conditioned reflexes around 1900. In Scandinavia, Johannsson and Tigerstedt developed the discipline of work physiology. In 1913, Rubner founded a work physiology institute in Germany. In the United States, Benedict and Cathcard described the efficiencies of muscular work in 1913, and the Harvard Fatigue Laboratory was established in the 1920s.

In the first half of the twentieth century, industrial physiology and psychology were well advanced and widely recognized, both in their theoretical research in studying human characteristics and in the application of this knowledge for the appropriate design of the living and working environment. Two distinct approaches to studying human characteristics had developed: one was concerned chiefly with the physiological and physical properties of humans, and the other focused mainly on the psychological and social traits. Although there was considerable overlap between these two approaches, it seems that in Europe, the physical and physiological aspects were more extensively studied, while in North America, experts focused more on psychological and social aspects.

EARLY DIRECTIONS IN EUROPE

Based on a broad foundation of anatomical, anthropological, and physiological research, applied or "work" physiology assumed great importance in Europe, especially during the hunger years associated with the First World War. Marginal living conditions stimulated research on the minimal nutrition required to perform certain physical activities; on the consumption of energy while carrying out agricultural, industrial, military, and household tasks; on the relationships between energy consumption and heart rate; on the assessment of muscular capabilities; on suitable body postures at work; on the design of equipment and workplaces to fit the human body; and on related topics. Another development in the 1920s was "psychotechnology," which involved testing individuals on their ability to perform physical and mental work, such as their vigilance and attention, their ability to carry mental workloads, their behavior as drivers of vehicles, and their ability to read road signs.

EARLY DIRECTIONS IN NORTH AMERICA

Most psychologists around 1900 were strictly scientific: they deliberately avoided studying problems that strayed outside the boundaries of pure research. Some investigators, however, had practical concerns such as sending and receiving Morse code, measuring perception and attention at work, using psychology in advertising, and promoting industrial efficiency. A particularly important step was the development of "intelligence testing" used to screen military recruits during the First World War and, later, to screen industrial workers for jobs deemed appropriate for their mental capabilities. The terms "intelligence testing" and "industrial psychology" won acceptance. (As an interesting aside, the term "industrial psychology" first appeared as a typographical error, actually meant to read *individual* psychology (Muchinsky, 2000). If the topic of intelligence testing intrigues you, please note that Gould (1981) provides a partly amusing, partly disturbing account of the early years of intelligence testing.)

NAMES FOR THE DISCIPLINE: "ERGONOMICS" AND "HUMAN FACTORS"

As mentioned earlier, the term "ergonomics" was first used in Poland in the late 1800s—and evidently it had virtually the same meaning that a group of British researchers had in mind when they met in Cambridge, on January 13 and 14, 1950, to give an encompassing name to their activities, which included anthropology, physiology, psychology, sociology, statistics, and engineering. All of these disciplines combine in studying humans at work, with the intent of using the information for the design of work tasks and equipment, and for the selection and training of workers. Apparently unwittingly—unaware of Jastrzebowski's earlier work—they reinvented the term "ergonomics," also deriving it from the Greek terms. Ergonomics was formally accepted as the name of the new society in 1950 (Edholm and Murrell, 1974; Monod and Valentin, 1979). This term is now used predominantly around the world; in the United States, the term "human factors" also came into play in 1956, with the word engineering often added or substituted to indicate application, such as in "human (factors) engineering" (Christensen et al., 1988). In 1992, the U.S. Human Factors Society renamed itself the Human Factors and Ergonomics Society.

There has been some discussion of whether human factors engineering differs from ergonomics—whether one relies more heavily on psychology or physiology, or is more theoretical or practical than the other. In essence, the two terms are synonymous; this is illustrated by the Canadian Society, which uses "human factors" in its English name and "ergonomie" in its French version.

TODAY'S ERGONOMIC KNOWLEDGE BASE

Under the pressure of the Second World War, the "human factor" as part of "man–machine systems" became a major concern. Technological development led to machines and systems that often put higher demands on the attention, strength, and endurance of individuals and teams than many could muster or tolerate. For example, in high-performance aircraft, the pilot was subjected to nearly unbearable accelerations; operators were required to monitor radar screens for hours at a time, tasked with detecting and distinguishing tiny blips from others. The development of space travel forced humans to fold themselves into miniscule capsules and to function in near weightlessness. Heavy physical work still persists in many jobs like commercial fishing and agricultural and forestry work—some are even new, such as in airline baggage handling. Many professional activities we all know today involve long hours using computers of one kind or another as the primary work tools.

Consequently, the field of ergonomics continues to grow and change, driven by new technologies and the resulting new tasks for people. Several classic sciences continue to provide the fundamental information we need to establish evolving ergonomic guidelines: anthropological understanding relies on anatomy, orthopedics, physiology, medicine, psychology, and sociology. Of course, physics, chemistry, statistics, and mathematical modeling also supply knowledge and methodology.

More applied disciplines have developed from these basic sciences; these include anthropometry, biomechanics, work physiology, industrial hygiene, management, and labor relations. The research areas overlap and intertwine; they produce practical ergonomic applications in industrial engineering, bioengineering, systems engineering, military engineering, and safety engineering—depicted in Figure A.1.

Ergonomists believe that "design begins with an understanding of the user's role in overall system performance and that systems exist to serve their users, whether they are consumers, system operators, production workers, or maintenance crews. This user-oriented design philosophy acknowledges human variability as a design parameter. The resultant designs incorporate features that take advantage of unique

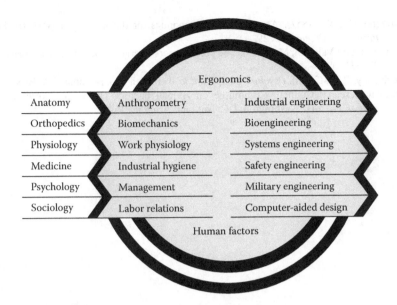

FIGURE A.1 Origins, developments, and applications of ergonomics. (Adapted from Kroemer et al. 2001.)

human capabilities as well as built-in safeguards" (National Research Council, 1983). The human here is the passenger, the participant, the operator, the supervisor, the controller, and the decision maker but, importantly, is never indentured to or victim of the system.

There is a hierarchy of goals in ergonomics. The most essential and basic task is to generate *tolerable* working conditions that do not pose unavoidable dangers to human life or health. When this basic requirement is assured, the next goal is to generate *acceptable* conditions to which the people involved can voluntarily agree. The final goal is to generate *optimal* conditions that are so well adapted to human characteristics, capabilities, and desires that physical, mental, and social well-being is achieved.

We define ergonomics (human factors, human engineering, human factors engineering) as "the study of human characteristics for the appropriate design of the living and work environment." Its fundamental aim is that all man-made tools, devices, equipment, machinery, and environments should advance, directly or indirectly, the safety, well-being, and performance of people.

REFERENCES

Christensen, J. M., Topmiller, D. A., and Gill, R. T. (1988). Human factors definitions revisited. *Human Factors Society Bulletin* 31, 7–8.

Edholm, O. G. and Murrell, K. H. F. (1974). *The Ergonomics Research Society. A History: 1949 to 1970.* Washington, DC: The Council of the Ergonomics Research Society.

Gould, S. J. (1981). *The Mismeasure of Man.* New York: Norton.

Jastrzebowski, W. (1857). An outline of ergonomics, or the science of work (in Polish). *Przyroda i Przemyst* 2, 29.

Monod, H. and Valentin, M. (1979). The predecessors of ergonomie (in French). *Ergonomics* 22, 673–680.

Muchinsky, P. M. (2000). *Psychology Applied to Work* (6th ed.). Belmont, CA: Brooks/Cole.

National Research Council, Committee on Human Factors. (1983). *Research Needs for Human Factors*. Washington, DC: National Academy Press.

Appendix B: A Brief History of Office Design

Human beings were not meant to sit in little cubicles staring at computer screens all day, filling out useless forms and listening to eight different bosses drone on about mission statements.

—**Peter Gibbons, character in movie *Office Space***

I always arrive late at the office, but I make up for it by leaving early.

—**Charles Lamb, writer**

Every year, we spend around 2000 hours working, and many of us spend those hours within the walls of a corporate space—a cubicle, perhaps, a shared desk in a conference room, or even a corner office. Given this massive chunk of time that we devote to work, our surroundings are certainly important to our well-being—yet, in the past, this was not a major consideration in office design.

Over the course of the past century, not only has our relationship with work evolved, but the spaces we occupy for work have changed as well. We can trace the history of office design through the key factors that influenced it: the tasks, the available technology, the prevailing management style, and the artistic vision of the architects, designers, and managers responsible for office spaces. Let us look at some of these factors.

TAYLORISM

In the early 1900s, office design was all about squeezing as much work as possible from employees, with minimal attention to employee comfort or morale. Such drudgery was perhaps prophesied by the word "office" itself, derived from the Latin for "duty." In the United States, the first commercial offices popped up in northern industrial cities like Chicago and New York. The engineer Frederick Taylor, credited as one of the first people to design an office space in the United States, placed many workers together in an "open environment."

The tasks in the workplace were repetitive and specialized; they mostly consisted of production-line clerical work. This fitted Taylor's expertise, because it meant that each worker became an expert in the repetitious task he or she was given and could perform efficiently and quickly, but doing so was monotonous and taxing. While workers sat in rows in large rooms and performed their tasks, uninterrupted work flows were supervised by managers. These bosses looked on from far more luxurious offices that surrounded the open space. While Taylorism—the term for this work style—stood for efficiency and output, it certainly did little or nothing for employee morale or job satisfaction.

The invention of the telegraph and the telephone—and subsequently electric lighting, typewriters, and calculating machines—spurred on the growth of "white-collar factories." Advances in mechanical engineering, particularly in steel frames and elevators, meant that buildings could grow taller and larger.

Well into the 1900s, the interiors of these large office buildings still followed the early example of the "Uffizi." Completed in 1581, Uffizi ("offices") in Florence, Italy, was built to accommodate the offices of the Florentine magistrates and now serves as a glittering gallery of Renaissance art. The building's design had a U-shape goundplan with an open court so that every room had a window providing some natural lighting and ventilation. When temperatures dropped, coal fires and stoves could be lit to warm the rooms, but the offices became malodorous and stuffy when windows had to be kept closed.

DEBUT OF AIR-CONDITIONING

Although some mechanical ventilation of the lower floors of tall office buildings had been in use since about 1890, design plans that followed shapes including U, O, E, I, H, and T were used into the 1930s so every room could have an outdoor window. In 1906, Frank Lloyd Wright's Larkin Administration Building was completed in Buffalo, New York, and it was a dramatic departure from the norm at the time. This structure took a significant step toward improving workers' environment; its central space was bathed in natural light and ringed with motivational inscriptions extolling the virtues of labor. It featured "sealed ventilation" to keep out pollution—trains passed by frequently—with early air-conditioning and radiant heat, stained glass windows, and even built-in office furniture; there were skylights and a light court, sound-absorbing treatments on cabinets, and a specially constructed floor that offered extra resilience. Similarly, Otto Wagner's Postsparkassenamt for Vienna, Austria, also built that year, took some notice of workers' plight; the main hall is designed like an atrium with a large glass skylight and plenty of windows throughout to allow for streams of natural light. The interior walls were non–weight bearing, so the interior offices could be rearranged as desired.

The Milam Building in San Antonio, Texas, built in 1936, was the first to be fully air-conditioned. Willis Carrier's high rise promised a 75°F, 56% relative humidity oasis in a southern clime known for oppressive heat and humidity, and this technology allowed for huge, city-block wide edifices with many rooms away from outside walls and windows (Arnold, 1999). Carrier, known as the "father of air-conditioning," helped usher in far more comfortable and efficient work conditions. The building housed not just office space but a post office, a barber shop, a drug store, and a restaurant; that place—called the Milam Grill—apparently became very popular, largely because of its "manufactured weather." Imagine throngs of lunch-goers, sweating and fanning themselves in the midday heat, eagerly standing in line to experience the comfort of refrigerated air within.

INTERNATIONAL STYLE

In Europe, at about the same time, the "International Style" was born and became popular; architects like Le Corbusier designed large office buildings that strayed from the customary "Beaux Arts" style. The International Style eschewed ornamental flourishes in favor of glass, steel, and concrete, with layout and design driven by logic. These buildings had Gustave Lyon's "regulated air" (l'air ponc-tuel) together with Le Corbusier's "neutralizing walls" (murs neutralisants) and cooled air circulating between double window panes. Espousing transparency and the "form follows function" philosophy, this style incorporated industrialized mass production techniques to create simplified and efficient designs.

In 1932, the PSFS building in Philadelphia was erected in the new style and was fully air-conditioned, which other contemporary high-rises—the Empire State and RCA buildings in New York—were not. Designed with a T-shaped floorplan that allowed a massive amount of natural light and rentable space, the tower soared from a huge base with a polished marble façade. The building incorporated the main characteristics of International Style architecture, with little ornamentation and an emphasis on functionality. The first floor featured retail space, the second housed the banking hall and related offices, and the top floor was reserved for executive suites and board meetings.

One edifice that really utilized all the new technology was Frank Lloyd Wright's Johnson Wax Administration Building in Racine, Wisconsin, completed in 1939. It was entirely sealed and fully air-conditioned. Its clerestory windows were con-structed from bundles of Pyrex glass tubing that provided diffuse lighting to inte-rior rooms. The large workspace ("Great Room") was beautifully lit with indirect light and very little glare, which resulted in an ideal environment for creative work. Wright deftly incorporated organic metaphors into his work; in the Johnson Wax Building, dendriform (treelike) columns rose 30 ft high and terminated at the roof level in broad, circular pads of concrete.

Air-conditioning was increasingly used in large office buildings, which, coupled with electric lighting, led to large and windowless buildings. After the Second World War, air-conditioning of large and small office buildings, even of single rooms by often-noisy window air conditioners, became commonplace in North America. While this allowed office dwellers to work in cooler climes, it also brought along new concerns: How did a lack of natural light affect health and performance? Was the quality of circulated air as good as outdoor air? Does overcooling or overheat-ing air artificially lead to lower work performance? And apart from physiological needs and effects, what about the psychological aspects? We are putting groups of individuals together in rooms, and they may not even like each other, but we are ask-ing that they get along, perform well, accept supervision, behave nicely, and—by the way—be happy. Is this realistic?

In the middle of the twentieth century, designers started to focus on the interior of the office itself, recognizing that office design was important, had an effect on the physical and psychological needs of workers, and consequently affected work output and productivity.

George Nelson created the first modern workstation in 1947. He felt strongly that designers should be constantly aware of the consequences their actions have on people and society, saying that "total design is nothing more or less than a process of relating everything to everything." Considering employees' physical and psycho-logical needs—and relating those met needs to increased productivity and output—marked a shift in office design, one that motivates office designers today.

In the 1950s in Europe, a socialist vibe prevailed as a type of revolt against Nazism, and this set of values descended on the office as well. Management was ousted from private office suites and made to mingle with the staff either via side-by-side workstations or other type of coworking arrangement to make communications more open and egalitarian. This *Bürolandschaft* (literally, office landscape) was intended to mirror and support the ebb and flow of human interactions in an "organic" way, with plants and carpets accenting and attenuating the rows of open office desks. The objective was to give workers an open type of office landscape to add flexibility and freedom while encouraging discussions and interactions.

Of course, offices during this time were increasingly noisy and chaotic, especially once typewriters and telephones joined the din. On the infrequent occasions where an office became hushed and silent, the too-quiet open space could appear unfriendly and preternaturally suppressed, with conversation nearly impossible as it would break the silence and become all too public. Add in the tension (at best) and all-out squabbles (at worst) that forced shared space brings with it—how warm or cool should the office air be, what to do about odorous food (microwave popcorn, we are looking at you), and personal radios filling others' air space—and *Bürolandschaft* quickly reveals some of its failings.

"ACTION" OFFICE SYSTEMS

In 1960, Robert Propst took a close look at workers' creativity and productivity. He interviewed workers, psychologists, and industrial relations experts, and he reviewed his own workstation and its drawbacks and advantages. By his analysis, he found the office environment sorely lacking. He stated that "today's office is a wasteland. It saps vitality, blocks talent, and frustrates accomplishment. It is the daily scene of unfulfilled intentions and failed effort." Ouch! He set out to redesign the office, beginning with modular desk units, including a stand-up desk—more on that in the following—and a walled writing desk. Calling it an "office system" when it debuted in 1964, this "action" system was a modular business furniture system, with low dividers and flexible work surfaces. This design is still around today, except we know it more familiarly as "cubicle."

About those stand-up desks, mentioned in the last paragraph: today (in 2015), stand-up desks are newly popular among health-conscious corporate employees, so some readers might be surprised that Propst's designs some 55 years ago were similar, in many respects, to the stand-up offices commonly used around 1900. The stand-up desk, for example, had a roll top, so that workers could safely leave their work out overnight, rather than clean them off every day before heading home. Probst's stand-up workstation also had a moveable display surface for source documents or magazines and a communications center stand that was insulated for better sound. Why did the stand-up desk not become popular and endure back then? Quite simply, supervisors did not like them because standing employees—and their output—are harder to observe than sitting employees.

Modular furnishings fulfilled a variety of functional requirements and allowed for a variety of differing arrangements. They "grew" when companies added staff and "shrunk" when companies contracted, and they allowed for various spatial plans depending on the task at hand in a given area. Modular furniture layout became the benchmark for a couple of decades, until it was taken to the extreme; at this point, it morphed into the dreaded cubicle farm.

CUBE FARMS

Around 1980, companies began to expand the layers of midlevel managers, and cubicles became the landing sites for this swelling rank of employees. Where to put this middle manager, who was too important for a simple exposed desk, but not yet elite enough for a window dwelling? You move them into a semiprivate, easily and cheaply configured, and deftly rearranged cubicle, of course, and *voila*! cubicle farms were born. Cubicles worked, but they did so somewhat inelegantly, because while cubicles replicated the furniture design and layout that Propst had developed, they did not incorporate the high-minded ideals that were behind his efforts. Other factors were in play at the time, and these factors all added up to make the cubicle increasingly unpopular. There was a change in the tax code, for example, which made it cheaper to set up an Action Office (cube) rather than a private office. Additionally, the series of crises the economy suffered in the 1970s was now followed by mergers and acquisitions, leveraged buyouts, and other chaotic and often traumatic realities. The cubicle, meant to give workers freedom and some autonomy, began to stand for boredom, frustration, and impermanence.

NETWORKING AND MIXED OFFICE DESIGN

The office continues to evolve during the first decades of the 2000s. With the tide turning against cubicle farms, furniture and office designers are shifting their focus to mixed-use spaces and "multitasking" furniture. The ideas of "modular" and "moveable"—the advantages of cubicles—seem to have staying power. Tables with low dividers, connected desks whose shape separates work areas, and semien-closed desks are all items we see in today's offices. Some demonstrate experimenta-tion with mixed-use spaces: certain areas in the building feature cubicles or private offices, while other areas contain semiprivate spaces or communal workstations that encourage coworking and collaboration. Still other spaces may be devoted to special uses—recreation and refreshments—and even outside spaces may be put to use by the office dwellers. For instance, there is a system called the Living Office; it incor-porates flowing, flexible design that fits the working style and specific tasks of a given organization and can quickly readapt to changing work environments.

Changing demographics are also exerting a substantial impact on the workplace. For example, younger workers—millenials—are accustomed to using wireless devices in casual settings where they can move around. They are not used to being tethered to workstations; they prefer adjustable workspaces. Consequently, in the world of office furniture, workstations that can convert from sitting to standing use are growing more and more popular.

Another reality of office design today is that many workers now work remotely, with one in five in the USA working from home, and that number is expected to grow

(Rapoza, 2013). This means that the office needs to deliver something special to be compelling enough for employees to actually decide to come in. And one important reason for some workers to come into the office space is to enjoy the social and collaborative aspect of colleagues and coworking. Consequently, today's office design needs to facilitate and even encourage collaboration and interaction.

What does the office of the future hold? Only time will tell, of course, but we envision wireless, moveable, interchangeable, multitasking, dynamic spaces that adapt constantly and evolve seamlessly. What do *you* think?

REFERENCES

Arnold, D. (1999). The evolution of modern office buildings and air conditioning. *ASHRAE Journal* 41, 40–54.

Baer, D. (2015). The origins of the awful open office layout. [Blog]. http://www.fastcompany. com/3007891/origins-awful-open-office-layout. Accessed November 10, 2015.

Del Castillo, M. (2015). An infographic history of office design. *Upstart Business Journal.* http://upstart.bizjournals.com/multimedia/interactives/2015/01/an-infographic-history-of-office-design-1920-to.html.

Guide to the Uffizi. (2015). http://www.uffizi.org/museum/uffizi-floor-plans/http://www. uffizi.org/museum/uffizi-floor-plans/.

Kesling, B. and Hagerty, J. (2013). Say goodbye to the office cubicle. *Wall Street Journal.* http://www.wsj.com/articles/SB10001424127887323466204578383022434680196.

Kroemer, K. H. E. (1997). *Ergonomic Design of Material Handling Systems.* Boca Raton, FL: CRC Press/Lewis.

McGregor, J. (2014). 9 things you didn't know about the office cubicle. *Washington Post.* https://www.washingtonpost.com/news/on-leadership/wp/2014/04/18/9-things-you-didnt-know-about-the-office-cubicle/.

Rapoza, K. (2013). One in five Americans work from home; numbers seen rising over 60%. *Forbes Magazine.* http://www.forbes.com/sites/kenrapoza/2013/02/18/one-in-five-americans-work-from-home-numbers-seen-rising-over-60/.

Remmele, M. (2012). How the office became what it is today. *Style Park.* http://www.stylepark. com/en/news/how-the-office-became-what-it.../335695.

Appendix C: Early Publications Related to Cumulative Trauma Disorders

Out of this nettle, danger, we pluck this flower, safety.

—**William Shakespeare**

We now have unshakable conviction that accident causes are man-made and that a man-made problem can be solved by men and women.

—**W.H. Cameron**

As mentioned throughout this text, keyboarding—whether done on a typewriter, a laptop, or a piano keyboard—can lead to an array of repetitive strain injuries, generally in keyboarders' hands and wrists. Cumulative trauma disorders (CTDs) are preventable, and this is what makes ergonomists cringe—if we can avoid these overexertion injuries, what is keeping us from doing so? While we offer a number of recommendations and guidelines for office ergonomics in this book, we show in this appendix that repetition-related injuries associated with keying have been covered in the literature not just over the past years and decades, but over centuries. Bernardino Ramazzini, often called the "Father of Industrial Hygiene," wrote about occupational repetitive injuries to office workers in the early 1700s (Wright, 1993). Of course, at that time, clerks generated records with ink and pen and suffered overexertion injuries related to writing, whereas repetition overexertions while keying on musical instruments became a medical concern about a century later. In the late 1800s, overuse disorders related to the operation of Morse devices appeared; this was followed by a flood of repetition injuries experienced by typists, as reported by Ayoub and Wittels (1989), Burnette and Ayoub (1989), Burry and Stoke (1985), Hochberg et al. (1983), Fry (1986a,b,c,d), Lockwood (1989), Pfeffer et al. (1988), and Schroetter (1925).

Tables C.1 through C.6 contain listings of related literature compiled from Kroemer's 2001 annotated bibliography that contains further commentaries and all reference citations.

TABLE C.1

Early Publications Related to Cumulative Trauma Disorders, Keying, and Keyboard Design

Year	Author(s)	Cumulative Trauma Disorder Aspects	Keying and Keyboard Design
1713	Ramazzini (Transl. Wright, 1993)	Diseases of workers due to "violent and irregular motions and unnatural postures of the body"	
1872	Poore	Discussion of the muscular causes for writers' cramp	Cramps common in pianists and telegraph operators
1868–1878	Sholes		Various keyboard designs shown in patents, including one—the last—with QWERTY layout
1887	Poore	Discussion of 21 cases of muscular syndromes in pianists	Overuse due to keying
1892	Osler	Continuous and excessive use of muscles in movements followed by spasm or cramp	Spasms and cramps common in pianists and telegraphists

Source: Adapted from Kroemer, K.H.E., *Int. J. Universal Access Inform. Soc.*, 1(2), 99, 2001.

TABLE C.2

Publications from 1900 to 1940 Related to Cumulative Trauma Disorders, Keyboarding, and Keyboard Design

Year	Author(s)	Cumulative Trauma Disorder Aspects	Keying and Keyboard Design
1909	Rowell		Proposal for new keyboard design to overcome known disadvantages
1915	Heidner		Patent of a split keyboard design to overcome motion and posture problems
1920	Banaji		Patent of a new key layout to overcome known disadvantages
1920	Nelson		Patent of a new key arrangement to overcome known disadvantages
1920	Wolcott		Patent of a new key layout design to overcome known disadvantages
1924	Hoke		Patent of a new split keyboard design to overcome motion problems
1924	Book		High motor ability of upper extremity necessary for champion typists
1925	Schroetter	Measurements of heart rate and oxygen consumption while typing.	
1926	Klockenberg	Repetitive key operation and posture cause serious problems in arms and hands.	Unhealthy posture while keying, due to keyboard design
1927	Zollinger	Seven cases of tendovaginitis after repetitive actions (929 cases of tendovaginitis mentioned).	
1930	Gilbert	Finger movements among keys are difficult.	Proposal for a new keyboard design to overcome motion problems
1930	Marloth		Patent of a split keyboard, with keys in rows perpendicular to natural positions of the forearms to achieve health and comfort for the typist
1931	Conn	Repetition-related tenosynovitis is a compensable occupational disease.	
1934	Biegel		Patent of a new keyboard design to avoid finger overloads
1934	Hammer	High-speed hand operations predispose for tenosynovitis.	
1936	Dvorak		Patent of a new keyboard design to avoid finger overloads

Source: Adapted from Kroemer, K.H.E., *Int. J. Universal Access Inform. Soc.*, 1(2), 99, 2001.

TABLE C.3

Publications from 1941 to 1960 Related to Cumulative Trauma Disorders, Keyboarding, and Keyboard Design

Year	Author(s)	Cumulative Trauma Disorder Aspects	Keying and Keyboard Design
1942	Flowerdew and Bode	Sixteen cases of tenosynovitis related to manual work	
1943	Dvorak	Mental tension and fatigue in typists with current keyboard design	New key assignments to avoid finger overloads
1949	Griffith		New key arrangement to achieve easier finger motions
1950	Hesh		Patent with fewer keys on the keyboard, movable keys
1951	Lundervold	Occupational myalgia in typists EMG recordings of involved muscles	
1954	Scales and Chapanis		Effects of keyboard slope on keying performance and subjective acceptance
1955	Hagen and Peters	Tenosynovitis as a result of repeated fast motions on the job	Most frequently observed in stenotypists
1956	Strong		Testing of Dvorak vs. standard keyboard
1957	Buckup	Tenosynovitis as a result of repeated motions	Associated with use of typewriters and other office machines
1958	Klemmer		Development and use of a 10-key typewriter keyboard
1958	Lundervold	EMG recordings of involved muscles	
1958	Hettinger	Test of disposition for tenosynovitis	
1959	Tanzer	Carpal tunnel syndrome (CTS) related to manual activities	CTS occurring in very busy secretarial work
1959	Lockhead and Klemmer		Testing of an 8-key typewriter keyboard
1960	Deininger		New key design features, force–displacement characteristics
1960	Creamer and Trumbo		Lateral declinations of the keyboard less fatiguing than horizontal arrangement
1960	Baader	Tendon, sheath, and muscle diseases due to frequent, repetitive, continued motions	Typists among afflicted workers

Source: Adapted from Kroemer, K.H.E., *Int. J. Universal Access Inform. Soc.*, 1(2), 99, 2001.

TABLE C.4

Publications from 1961 to 1980 Related to Cumulative Trauma Disorders, Keyboarding, and Keyboard Design

Year	Author(s)	Cumulative Trauma Disorder Aspects	Keying and Keyboard Design
1961	Droege and Hill		Performance on electric and on mechanical typewriters.
1961	Peres	Repetitive activities in hand and wrist result in strain, pain, and cumulative injuries.	
1961	Ratz and Ritchie		In keyboarding, motor system constraints predominate over choice reaction time.
1962	Diehl and Seibel		Removal of visual and particularly auditory feedback reduces typing performance.
1962	Klemmer and Lockhead		Between 56,000 and 83,000 keystrokes per day are performed by IBM-machine operators.
1962	Seibel		On chord keyboard, increased performance is achieved by decreased response times.
1962	Seibel and Rochester		Keys rearranged to reduce finger movements and effort.
1962	Yllo		Right-hand keyboard was rearranged to avoid unnatural wrist and elbow angle. Lowering keyboard and tilting it down to the right side reduced fatigue as determined by EMG.
1962	IBM		Internal proposal for new keyboard design to overcome known disadvantages.
1963	Robbins	CTS related to wrist flexion and extension.	
1963	Cornog et al.		Chord keyboard feasible for address encoding.
1963	Seibel		Response times increase with number of alternatives. Reaction time on keys increases with information size.
1963	Morgan et al.		Recommendations for selection and design of controls.
1964a,b	Kroemer (two publications)	CTDs related to force, displacement of keys, frequency, and wrist posture while keying.	New keyboard design: split keyboard, tilt each half laterally, and rearrange keys.

(Continued)

TABLE C.4 (*Continued*)

Publications from 1961 to 1980 Related to Cumulative Trauma Disorders, Keyboarding, and Keyboard Design

Year	Author(s)	Cumulative Trauma Disorder Aspects	Keying and Keyboard Design
1964	Seibel		Final keying performance cannot be predicted but must be observed.
1964	Fox and Stansfield		Keyboard design determines keying rate.
1965	Bowen and Guinness		Keyboards should be designed rationally and in keeping with good human engineering practice. On such keyboards, performance differences are small.
1965	Conrad and Longman		Keying feedback important while learning.
1965	Galitz		Keying force, travel, activation, slope, and feedback.
1965	Kroemer	Typists terminated keying because of aches and complaints in arms, wrists, and fingers.	Author proposes tilted rather than horizontal keyboards; tilt postponed fatigue.
1965	Mayzner and Tresselt		Single-letter and diagram frequencies.
1966a	Galitz	Fatigue associated with keying.	Key force, displacement, and slope.
1966b	Galitz		Key force and activation feedback on microswitch keys are important.
1966	Phalen	754 cases of CTS are related to finger flexion with wrist flexion, with references to numerous reports.	
1966	Hymovich and Lindholm	62 cases of CTDs with repetitive work; more reported in females than in males.	
1966	Tichauer	"…modern industrial health care must consider… impairment caused insidiously…by gradual, cumulative, and often imperceptible overstrain of minute body elements."	Attention must be paid to the hand–tool interface.
1967	Max-Planck Gesellschaft	Muscular fatigue of keyboard operators can be reduced by proper design.	Patent of a split and tilted keyboard with hand-configured key arrangements.
1967	Lewin		Pointing pen/electronic keyboard as input device.
1967	Smith		Mistakes made in data entry on key sets.

(*Continued*)

TABLE C.4 (*Continued*)

Publications from 1961 to 1980 Related to Cumulative Trauma Disorders, Keyboarding, and Keyboard Design

Year	Author(s)	Cumulative Trauma Disorder Aspects	Keying and Keyboard Design
1967	West		Deprivation of visual feedback while typing increases errors.
1967	Ulich	Use of EMG to assess muscular activities.	
1968	Lhose		Alphanumeric keyboard standard.
1968	Adams		Suitable feedback is important for key operation.
1968	Keele		Theories of movement control and the importance of feedback.
1969	Kinkead and Gonzales		Effects of key designs, including tactile feedback.
1969	Remington and Rogers		Listing of more than 300 publications on keyboard and keying collected by IBM.
1969	West	Fatigue and performance during 30 minute typing sessions.	
1970	Caldwell	Muscle fatigue after repeated exertions.	
1970	Galitz and Laska		Manual activities of computer operators.
1970	Hirsch		Standard vs. alphabetic keyboard.
1971	Garrett		Hand dimensions and biomechanical characteristics.
1971	Klemmer		Summary of 35 articles regarding keyboard design, feedback, key force, and displacement.
1971	Komoike and Horiguchi	233 fatigue cases in female office workers: effects of high speed and paced rhythm, localized fatigue, and pain.	
1971	Phillips and Kincaid		Better key arrangement is needed and technically easy.
1971	Michaels		QWERTY keyboard allows better performance than alphabetic keyboards.
1971	Smith and Goodwin		Alphabetic data entry by telephone key arrangement possible.
1972	Gallaway		Recommendations for key force and displacement.
1972	Kroemer	Effects of key operation on operator strain and performance.	Keyboard arrangement, tilted to the side.

(Continued)

TABLE C.4 (*Continued*)
Publications from 1961 to 1980 Related to Cumulative Trauma Disorders, Keyboarding, and Keyboard Design

Year	Author(s)	Cumulative Trauma Disorder Aspects	Keying and Keyboard Design
1972	Phalen	598 cases of CTD.	
1972	Samuel		New key designs.
1972	Seibel		Key force, displacement, and feedback.
1972	Smith		Alternative function keys.
1972	Alden et al.	Hand musculature limits keying and fatigue, key force, and displacement.	Review of, and recommendations for, keyboard design, key force and displacement, and feedback.
1973	Tichauer	Cites Ramazzini's 1713 statement about diseases due to violent and irregular motions.	
1973	Yoder et al.	Muscle fatigue and wrist problems in assembly, assessment by EMG.	
1974	Duncan and Ferguson	Occupational cramps and myalgia associated with adverse postures of keyboarders; ninety patients with cramps and myalgia in keyboard operation; effects of keyboard design and arrangement on posture.	New key and keyboard arrangement.
1974	Ferguson and Duncan	Adverse postures of digits, wrists, arms, shoulders, neck, and trunk due to keyboard layout and arrangement.	Key arrangement, division of keyboards for each hand, tilted sideways.
1974	Kassab	Wrist support reduces muscle effort as assessed by EMG.	Wrist support reduces muscle effort as assessed by EMG.
1974	Herndon et al.	Reference to previous publications on CTDs.	
1974	Maeda	Disorders observed in business machine operators: effects of postures and motions.	
1974	Showel		Typist training methods.
1974	Sidorsky		Alpha-dot keyboard developed and tested.
1975	Goodwin		Cursor control keys.
1975	Ayoub et al.	Changes in low frequency of the EMG as measure of muscle fatigue.	
1975	Hanes		Design guidelines for keyboards and keys.
1975	Einbinder		Patent to minimize finger motions by rearranging keys, curved key rows, slanted key tops, and split keyboard.
1975	Engel and Granda		Cursor control devices, including keyboard.

(*Continued*)

TABLE C.4 (*Continued*)

Publications from 1961 to 1980 Related to Cumulative Trauma Disorders, Keyboarding, and Keyboard Design

Year	Author(s)	Cumulative Trauma Disorder Aspects	Keying and Keyboard Design
1975	Birkbeck and Beer	Light highly repetitive finger and wrist motions causal factor for CTS.	High incidence of cashiers/ secretaries among CTS patients.
1976	Clare		Resistance change during key travel to indicate activation (tactile feedback).
1976	Wood		Ergonomic console design.
1976	Davies and Pratt	Effects of muscle contraction on manipulation.	
1976	Ferguson and Duncan	29 CTD patients treated.	Keyboard design and operating posture associated with CTDs.
1976	Posch and Marcotte	Of 1201 cases of CTD, 36% were work related.	
1976	Einbinder		Patent to minimize finger motions by rearranging keys, curved key rows, slanted key tops, and split keyboard.
1976	Tichauer	EMG and biomechanical procedures available to relate muscular effort and motion.	
1976	Van Nes		Among keying errors, activating an adjacent (incorrect) key was the second or third most often recording error of seven error categories.
1977	Hadler	CTD may be precipitated or caused by stereotyped hand use, but little formal proof available.	
1977	LeCocq		Ergonomic design of computer terminal.
1977	Bequaert and Rochester		Experiments on a chord keyboard.
1977	Rodbard and Weiss		Effects of reduced blood supply in arms of typists.
1977	Roth et al.		Ergonomic console design, specifically for reach.
1978	Dorris and Purswell	Warnings needed because people are not proficient at assessing hazards.	
1978	Card et al.		Ergonomic evaluation of input devices including keys.
1978a,b	Rochester et al. (two publications)		Chord keyboard developed.
1978	Whitaker		Moveable keys contain an arrangement to follow the fingertip positions.

(*Continued*)

TABLE C.4 (*Continued*)

Publications from 1961 to 1980 Related to Cumulative Trauma Disorders, Keyboarding, and Keyboard Design

Year	Author(s)	Cumulative Trauma Disorder Aspects	Keying and Keyboard Design
1978	Zapp		Keys moveable by tips of the finger, provision of hand/wrist supports.
1978	Rothfleisch and Sherman	Occupational risk of CTS can be reduced by proper biomechanical and ergonomic means at the job.	
1978	Tichauer and Gage	Tenosynovitis and hand overuse.	
1978	Armstrong and Chaffin	Displacements of wrist and finger flexor tendons.	Displacements of wrist and finger flexor tendons.
1979	Cakir et al.		Details for keyboards and computer workstations in general.
1979a, b	Armstrong and Chaffin (two publications)	18 cases of CTD; effects of force and wrist positions.	
1979	Gopher and Eilam		New chord keyboard.
1979	Kuorinka and Koskinen	17 cases of wrist syndromes related to number of work projects handled.	
1979	Luopajaervi et al.	In 152 female assembly workers, CTDs were more prevalent than in shop assistants.	
1979	Waris et al.	Methods to screen for CTDs.	
1980	Huenting et al.	CTDs frequent in accounting machine operators.	CTDs related to posture and key operation.
1980	Maeda et al.	CTDs found more in keyboard operators than in sales clerks.	Recommend improvements in keyboard operator workstations, including arm support.
1980	Cakir et al.		Details for keyboards and computer workstations in general.
1980	Laeubli et al.	Same findings in Switzerland as in Japan; occupational CTDs related to work with office machines.	Keyboard height related to body posture.
1980	Zapp		Keys arranged in hand-configured array, wrist rest provided.

Source: Adapted from Kroemer, K.H.E., *Int. J. Universal Access Inform. Soc.*, 1(2), 99, 2001.

TABLE C.5
Publications from 1981 to 1990 Related to Cumulative Trauma Disorders, Keyboarding, and Keyboard Design

Year	Author(s)	Cumulative Trauma Disorder Aspects	Keying and Keyboard Design
1981	Cannon et al.	CTDs related to jobs.	
1981	Grandjean et al.		Split, slanted, and tilted keyboard preferred and associated with natural postures.
1981	Fraser	Tenosynovitis related to excessive and repeated motions in keyboard operation; disorders due to overuse, especially excessive key force and displacement, and keying frequency.	Body posture at work, related to workstation layout.
1981	IBM		Ergonomic guide for managers of VDT workplaces.
1981	IBM		Ergonomics of VDT workstations.
1981	Zipp et al.		EMG findings suggest slant and tilt of a split keyboard.
1981	Litterick		Review of keyboard improvements.
1981	Smith et al.	Musculoskeletal problems more often found with VDT operators than with control subjects.	
1981	Stammerjohn et al.		Assessment of physical working conditions, particularly of keyboard height and chair.
1981	Miller and Suther		Preferred adjustments for keyboard height, keyboard slope, and the VDT.
1981	Tynan		A quick and easy procedure to determine ergonomic design aspects of a computer workstation.
1981	Dainoff	Health issues associated with work at VDTs.	Ergonomic issues associated with work at VDTs.
1981	Miller and Suther		Preferred height and angle settings of display and keyboard.
1981	Happ and Beaver	Strong association between fatigue and visual stress suggests the same construct as for musculoskeletal complaints.	
1981	Benz et al.		Comprehensive ergonomic guide for VDT workplaces, including keyboard and key design.

(Continued)

TABLE C.5 (*Continued*)

Publications from 1981 to 1990 Related to Cumulative Trauma Disorders, Keyboarding, and Keyboard Design

Year	Author(s)	Cumulative Trauma Disorder Aspects	Keying and Keyboard Design
1981	Cannon et al.	CTS associated with use of vibrating tools.	
1981	Eikelberger		Current technology allows easy custom design of keyboards, including the use of keys with "snap action" for tactile feedback.
1981	Gentner		Study of finger movements while typing.
1981	Hirsch		Examples of IBM human factors research on keyboards, including feedback.
1981	Malt		Keys arranged for frequency of occurrence. Keys are on concave surface; keyboard is split.
1981	Simonelli		The arrangement of keys on a membrane keypad influenced keying time.
1982	Grandjean		Postural problems related to VDU workplaces.
1982	Huenting et al.		Constrained postures may be associated with physical impairments. The incidence of complaints is lowered if hands and forearms can be rested.
1982	McPhee	Incidence of repetition injuries increasing in Australia; RSI related to frequency, force, posture, and time.	
1982	Cohen		Typing performance with membrane keyboard is nearly as good with conventional keyboard.
1982	Helander		Human factors guidelines for visual display terminals.
1982	Norman and Fisher		Lack of justification for current keyboard size and layout and radical redesign of the keyboard promising.
1982	Nakaseko et al.		Split keyboard, laterally tilted, and sections slanted improve posture and reduce fatigue.

(*Continued*)

TABLE C.5 (*Continued*)

Publications from 1981 to 1990 Related to Cumulative Trauma Disorders, Keyboarding, and Keyboard Design

Year	Author(s)	Cumulative Trauma Disorder Aspects	Keying and Keyboard Design
1982	Emmons and Hirsch		Comparison of keyboards of 5°–18° slope.
1982	Dainoff	Visual and musculoskeletal complaints among VDT operators are high.	Linkages among stress symptoms and ergonomics of VDT workstations.
1982	Suther and McTyre		Keyboard slopes between 10° and 25° are recommended.
1982	Armstrong et al.	CTDs can be caused, precipitated, or aggravated by repeated exertions with the hand.	
1982	Wagner		Review of the European requirements and recommendations for NCR keyboards and their use.
1982	Haider et al.		Ergonomic design of keyboarding workplaces, keyboards, and keys.
1982	Helander		Office workstation design.
1982	Lundborg et al.	Effects of ischemia and wrist compression on nerve function.	
1982	Butterbaugh		Four key layouts had equally accurate inputs.
1982	Price et al.	Operator performance measures include muscular fatigue.	Body posture changes with duration of typing.
1983	Armstrong	CTD is an occupational illness.	Design tools and tasks so that wrist displacements are avoided.
1983	Jensen et al.	In 1979, more than 3000 workers' compensation claims reported for nonimpact wrist disorders in 26 U.S. states, constituting 6% of all compensable cases.	
1983a	Noyes		Review of the history of QWERTY keyboard; proposals to change the design.
1983b	Noyes		Characteristics, advantages, and disadvantages of chord keyboards vs. sequential keyboards.
1983	Najjar		Review of 20 publications on keyboards since 1967.

(*Continued*)

TABLE C.5 (*Continued*)
Publications from 1981 to 1990 Related to Cumulative Trauma Disorders, Keyboarding, and Keyboard Design

Year	Author(s)	Cumulative Trauma Disorder Aspects	Keying and Keyboard Design
1983	AT&T Bell Laboratories		A comprehensive guide for VDT workstation ergonomics, including keyboards.
1983	Clarke and Caroll		How to write user-friendly manuals.
1983	Murray et al.		Voice vs. keyboard control of cursor in work processing.
1983	Arndt	Long recognized that repetitive motion may lead to CTDs.	Posture-related complaints at VDT workstations.
1983	Miller and Suther		Preferred height settings of keyboards.
1983	Grandjean et al.	Reduced complaints with comfortable postures in adjustable workstations.	Ability to adjust working posture; effects of adjustable furniture on work postures.
1983a,b,c	Kroemer (three publications)		Ergonomics of VDT workplaces.
1983	Kaplan	Typing as an occupation may contribute to the development of carpal tunnel syndrome.	
1983	Chisvin	VDT operators experience more physical discomfort than other clerical employees, particularly in the neck and shoulders.	Postures at VDT work.
1983	Gopher and Koenig		Development of a chord keyboard.
1983	Benz et al.		Comprehensive ergonomic guide for VDU workplaces, including keyboard and key design.
1983	Monty et al.		Acoustic feedback of key operation on several keyboards resulted in faster entry and was strongly preferred
1983	Francas et al.		Comparison of key sets for alphabetic entry.
1983	Hochberg et al.	CTDs among musicians.	
1983	Zipp et al.	CTDs frequent among keyboarders.	Ergonomically designed keyboard recommended, split, tilted, and slanted.
1983	Snyder		Recommendations for keys and keyboard: height, slope, key placement, and feedback.

(*Continued*)

TABLE C.5 (*Continued*)

Publications from 1981 to 1990 Related to Cumulative Trauma Disorders, Keyboarding, and Keyboard Design

Year	Author(s)	Cumulative Trauma Disorder Aspects	Keying and Keyboard Design
1984	NIOSH	VDT operators report musculoskeletal strains and discomfort.	Improving ergonomic conditions reduces musculoskeletal complaints.
1984	Peters		15 cardinal warning principles.
1984	Burke et al.		Effects of keyboard height on performance and preference.
1984	Joyce	Behavioral training resulted in reduced back, neck, and shoulder pains. Increased occurrences of stiff/sore wrists and loss of feeling in wrists/fingers with increased performance.	
1984	Brunner and Richardson		Tactile feedback about key activation improves performance and acceptance.
1984	Salthouse		High-keying rates depend on perceptual chunking and simultaneous execution of digit motions.
1984	Gopher et al.		High input performance on a chord keyboard can be achieved by proper spatial key arrangements and considerations of hand symmetry.
1984	Buesen		Development of a split keyboard with tilted halves.
1984	Rosch		Comparison of existing keyboards in their differences in tactile feedback.
1984	Life and Pheasant		Higher keyboards related to muscular effort.
1984	Grandjean et al.	Preferred settings of adjustable workstations reduce complaints in the neck, shoulder, and back.	If adjustable, different operators set the keyboard height to widely varying levels.
1984	Shute and Starr		Discomfort may be reduced with totally adjustable workstation equipment.
1984	Starr		VDT use has the same physical discomfort as, but better job satisfaction than, use of paper documents for the same job.

(Continued)

TABLE C.5 (*Continued*)
Publications from 1981 to 1990 Related to Cumulative Trauma Disorders, Keyboarding, and Keyboard Design

Year	Author(s)	Cumulative Trauma Disorder Aspects	Keying and Keyboard Design
1984	Thomas		The effects of ZH 1/618 requirements on VDT offices.
1984	Cumming		Software advances make key-character assignments changeable.
1984	Helander et al.	A comprehensive review of 82 studies concerning work with VDTs.	A comprehensive review of 82 studies concerning work with VDTs.
1984	Arndt		Key, keyboard, and workstation design.
1984	Browne et al.	RSI defined as injuries caused by overload from repeated use or by maintaining constrained postures.	Keyboard operators at risk.
1984	Hadler	Questions CTS and job association. Lists hand and wrist disorders felt to be related to overuse.	
1984	Arndt et al.	NIOSH manual for a course on health issues for VDT supervisors.	NIOSH manual for a course on health issues for VDT supervisors.
1984	Frey et al.		Keyboard is split into left and right halves, slanted, and tilted down to the sides. Additional sets of keys are arranged further on each side, but not tiled. All key fields are located on planes sloping up away from the operator who uses the slopes to rest the arms.
1985	Louis	Fatigue with conventional keyboard.	Patent of a split keyboard.
1985	Gaydos	Discussion of VDT issues before a subcommittee of the House of Representatives.	Discussion of VDT issues before a subcommittee of the U.S. House of Representatives.
1985	Koppa		Review of keypad layouts.
1985a	Bartram and Feggou		Key activation affected by finger strength and mobility.
1985	Nakaseko et al.	Hand posture, depending upon keyboard arrangement, affects musculoskeletal complaints.	A split keyboard, both slanted in the top view and tilted in the front view, is preferred over regular keyboard.
1985	Gopher et al.		Investigation of coding and arrangements of chord keyboards.

(Continued)

TABLE C.5 (*Continued*)
Publications from 1981 to 1990 Related to Cumulative Trauma Disorders,
Keyboarding, and Keyboard Design

Year	Author(s)	Cumulative Trauma Disorder Aspects	Keying and Keyboard Design
1985	Lynch		ANSI Standard on VDT workstations in the final stage of acceptance as an American National Standard.
1985	Thomas		Promotion of good human factors in IBM products.
1985	Stewart	Fatigue and discomfort from awkward postures caused by poorly located controls.	
1985	Oxenburgh et al.	Duration of keyboarding and job organization are related to CTD likelihood, as are psychosocial factors.	
1985b	Bartram and Feggou		Keying performance depends on keyboard design.
1985	Westgaard and Araas		Ergonomic improvements at work reduce CTD occurrence.
1985	McKenzie et al.	Successful industry program to control CTDs based on better hand–tool design, training, and management.	
1985	Knave et al.	Discomfort with VDT work.	
1985	Burry and Stoke	Repetitive strain injuries of muscles and tendons in various body parts. Among the three main causes are poorly designed workstations and long periods of repetitive work.	Among the three main causes of RSI are poorly designed workstations and long periods of repetitive work.
1985	Brown et al.	The Hettinger test is able to predict whether individuals are susceptible to RSI.	
1986	Kirschenbaum et al.		Chorded keyboard designed to minimize physical exertion.
1986	Gilad and Pollatschek		Simulation tool for keyboard layout.
1986	Gopher		Two-hand keyboard developed to provide alternative to standard keyboard.
1986	Silverstein et al.	Prevention strategies to avoid CTS.	

(*Continued*)

TABLE C.5 (*Continued*)
Publications from 1981 to 1990 Related to Cumulative Trauma Disorders, Keyboarding, and Keyboard Design

Year	Author(s)	Cumulative Trauma Disorder Aspects	Keying and Keyboard Design
1986	Rosch		New keyboard designs to replace conventional keyboards.
1986	Standard Telephone and Radio AG, ITT		Use of the ergonomic STR keyboard.
1986	Hagberg and Sundalin	EMG used to assess muscle use and discomfort in word processor operators.	
1986	Armstrong	CTD risk factors include repetitive exertions and body postures.	
1986	Armstrong, Radwin et al.	Repetitiveness, force, mechanical stresses, posture, vibration, and temperature are CTD factors.	
1986	Fine et al.	Surveillance efforts to determine causal factors of CTDs.	
1986	Sauter et al.	CTDs related to keyboard use, wrist displacements, lack of wrist support, or pressure on edges.	Recommendations for keyboard placement and use of wrist rests.
1986	Marsh	Instrument for testing sensibility of CTS patients.	
1986	Statshaelsan		Detailed ergonomic recommendations for keys and keyboards.
1986a, b, c, d	Fry (four publications)	Overuse syndrome in musicians extensively reported from 1830, with 75 publications between 1830 and 1911 reviewed. Physical signs in hand and wrist associated with overuse injury in musicians.	
1986	Armstrong et al.	Repetitive motions, forceful motions, and posture are CTD risks.	Keyboard use exposes operators to reported CTD factors such as repetitive motions and postures.
1986	Hodges		Patent to avoid the unnatural positions of arms and wrists that are required at the conventional keyboard; keys and keysets can be individually adjusted to achieve normal, natural, or restful positions for the human hands.

(Continued)

TABLE C.5 (*Continued*)

Publications from 1981 to 1990 Related to Cumulative Trauma Disorders, Keyboarding, and Keyboard Design

Year	Author(s)	Cumulative Trauma Disorder Aspects	Keying and Keyboard Design
1986	Pollatschek and Gilad		To overcome the biomechanical disadvantages of the QWERTY keyboard, a means to customize any keyboard cheaply and ineffectively is presented.
1986	Seligman et al.	NIOSH found CTS associated with wrist and hand posture during typing.	Adjust keyboard height relative to seat; provide wrist rests.
1986	Smith	Keying can produce tenosynovitis and carpal tunnel syndrome. Musculoskeletal problems foremost health concern of VDT use.	
1987	Chatterjee	CTD review, risk of injury both occupational (at work) and at leisure.	Recommendations for keying work design.
1987	Hocking	More than 2200 RSI reports in about 5000 clerical positions in Australia in the early 1980s.	
1987	Ferguson	RSI epidemic in 1970s and 1980s in Australia.	
1987	Richardson et al.		Investigation of various keyboards.
1987a	Silverstein et al.	CTS associated with force and repetition at work.	
1987	Wiklund et al.		Keyboard comparisons.
1987a	Grandjean		Design of keys and keyboards.
1987	Raij et al.		Motor and perceptual implications of chord keying.
1987	Armstrong et al.	Relationship between repetitiveness and forcefulness of manual work and biomechanical factors in tendonitis.	
1987	Greenstein and Arnaut		Recommendations for key displacement, key force, and keying feedback.
1987	Blair and Bear-Lehman	Emergence of CTDs like an epidemic.	
1987	Louis	Course of CTD, and its treatment, predictable.	
1987b	Silverstein et al.	Plant workers with hand–wrist CTD tend to transfer out of their jobs.	

(*Continued*)

TABLE C.5 (*Continued*)
Publications from 1981 to 1990 Related to Cumulative Trauma Disorders, Keyboarding, and Keyboard Design

Year	Author(s)	Cumulative Trauma Disorder Aspects	Keying and Keyboard Design
1987	Bleecker	The cross-sectional size of the carpal tunnel may be a risk factor.	
1987	Feldman et al.	Surveillance and ergonomic intervention to reduce the risk of CTS in industry.	
1987	Gomer et al.	EMG is a useful tool to assess muscular activities and fatigue in keyboard operation.	
1987	Rossignol et al.	Increased prevalence of musculoskeletal conditions in work with VDTs, apparently related to the duration of VDT work.	
1987	Seror	Tinel's sign as diagnostic for CTS questioned.	
1987	Sauter et al.	Pressure at edge of wrist may cause trauma.	Wrist support should not present sharp pressure point, but be gently contoured and padded.
1987b	Grandjean		Recommendations for VDT keyboards and workstations.
1987	Herzog and Herzog		Patent of keys and keyboard sections aligned to provide proper posture of forearms and hands and to provide accurate and unobstructed and comfortable movements of the fingers.
1987	Kiser	Low-load hand activities can result in significant stress on muscles and tendons, especially with the wrist not in the straight position.	
1987	Goldstein et al.	Accumulation of strain occurs in human flexor digitorum profundus tissues.	
1988	Patkin	CTDs in hand–arm region.	
1988	Kroemer		Survey of, and recommendations for, ergonomic means to computer workstation design.
1988	Pfeffer et al.	In 1960, CTD most frequently diagnosed, best understood, and most easily treated entrapment neuropathy.	

(*Continued*)

TABLE C.5 (*Continued*)
Publications from 1981 to 1990 Related to Cumulative Trauma Disorders, Keyboarding, and Keyboard Design

Year	Author(s)	Cumulative Trauma Disorder Aspects	Keying and Keyboard Design
1988	Kiesler and Finholt	History of Australian RSI.	
1988	Foster and Frye	RSI is a case of conflicts in medical knowledge.	
1988	Gopher and Raij		For mainly cognitive reasons, entry performance on separate and vertical chord keyboards for each hand was significantly faster than at a standard QWERTY layout.
1988	Carr		Forces applied to keys depend on their locations.
1988	Human Factors Society ANSI/HFS 100-1988		Standards set for keys: travel 1.5–6 mm, preferred 2.0–4.0 mm. Force 0.25–1.5 N (preferred 0.5–0.6 N); feedback tactile or auditory, or both (tactile feedback preferred); reduction in key force after about 40% of total displacement.
1988	Rose	To avoid muscle overuse problems, China should learn from our lessons and use accumulated ergonomic knowledge for the design of a keyboard for the Chinese language.	On a keyboard for the Chinese language, the maximum number of keys should be approximately 44. All key columns should be aligned rather than staggered. Most frequently used keys should be located on the home row, possibly operated by the thumb. Hand pronation and lateral wrist deviations should be avoided through proper key set arrangement. Keyboard should be split and arranged according to existing designs in Germany, Switzerland, and Great Britain.
1988	Baidya and Stevenson	EMG measurements to assess fatigue associated with wrist extensions.	

(Continued)

TABLE C.5 (*Continued*)
Publications from 1981 to 1990 Related to Cumulative Trauma Disorders, Keyboarding, and Keyboard Design

Year	Author(s)	Cumulative Trauma Disorder Aspects	Keying and Keyboard Design
1988	Hobday		To avoid lateral wrist deviation and pronation and to follow hand shape and finger motions, the keyboard is divided into separate keysets, with the keys tilted and arranged on arcs.
1988	Green et al.	Variability of joint locations.	Wide range of preferred postures of keyboard operators.
1988	Bjoerksten	High occurrence of neck, shoulder, and low back problems among secretaries, partly explained from EMGs.	
1988	Molan and Sikorski	Persons with and without CTDs showed similar work behavior.	
1988	Erdelyi et al.		EMG measurements to determine suitable work postures.
1988	Itani et al.	EMG suitable means for determining muscular activities while keyboarding.	
1988	Krueger et al.	Repetitive work is one reason for musculoskeletal complaints.	
1988	Nathan et al.	No consistent association between occupational activity and slowed conduction in median nerve found.	
1988	Langley		Patent of a chord keyboard that can be operated without movement of fingers from one key to another. Fingers can rest on keys without activating them.
1988	Putz-Anderson	Manual describes and defines CTDs, especially of the upper extremities.	Management and engineering methods used in combination "to make the job fit the person, not to make the person fit the job" by redesigning the job or the tool to reduce the job demands of force, repetition, and posture.
1989a	Burnette and Ayoub	Definition of CTDs, prevalence, costs, pathology and etiology, treatment and rehabilitation, and prevention.	

(*Continued*)

TABLE C.5 (*Continued*)
Publications from 1981 to 1990 Related to Cumulative Trauma Disorders,
Keyboarding, and Keyboard Design

Year	Author(s)	Cumulative Trauma Disorder Aspects	Keying and Keyboard Design
1989b	Burnette and Ayoub	Computerized model to determine the CTD risk as a function of job stress and moderating factors.	
1989	Ayoub and Wittels	Description of CTD injury mechanics.	Workplace design, education and training, and supervisory and managerial contributions are all important.
1989	Hadler	No impressive data to support the contention that any upper extremity usage, within reason, is damaging.	
1989	Center for Office Technology		Reports on and summarizes 21 publications on musculoskeletal research and resulting workstation design recommendations, including keyboards.
1989	Morita		Keyboard design for the Japanese language with the keys divided into groups for the left and right hand, with keys on the outside located lower than those in the center. Keys arranged for ease of motion and reduction of movements, all for fast operation without excess fatigue.
1989	Knight and Retter		New radically different key entry device in which the keyboard is split and where the fingertip operates switches in different directions, from the same location. This avoids bending the wrist, flattening the hand, keeping the hands in position, repeated similar strain, and energy to operate keys.
1989a	Green and Briggs		Operators must be trained in correct use.

(*Continued*)

TABLE C.5 (*Continued*)
Publications from 1981 to 1990 Related to Cumulative Trauma Disorders, Keyboarding, and Keyboard Design

Year	Author(s)	Cumulative Trauma Disorder Aspects	Keying and Keyboard Design
1989b	Green and Briggs	No differences between anthropometry of male CTD sufferers and nonsufferers; in contrast, significant anthropometric differences among female groups of sufferers and nonsufferers.	
1989	Lockwood	CTD problems in musicians well known.	
1989	Kroemer	Types and possible/likely causes of CTDs; ergonomic interventions to avoid CTDs.	
1989	Williams et al.	Exercises did not result in improvements regarding CTS.	
1989	Sind		For membrane keyboards, recommends lowest possible activation force and feedback about actuation.
1989	National Occupational. Health and Safety Commission (Australia)	Organizational and design means to prevent overuse syndromes associated with keyboarding.	Design guidance to prevent overuse syndromes associated with keyboarding.
1989	Rempel et al.	Wrist tendonitis and CTS were common complaints among VDT users whose workplaces showed common ergonomic problems.	Recommendations for VDT workstations.
1990	Zapp		Patent of new key design; the keys are tilted instead of tapped; wrist rest is provided.
1990	Faubert and Pritchard	CTDs occur in keyboard users.	Reposition keys, reshape keyboard, and consider flexible or hinged keyboards. Provide alternate keyboards.
1990	Hughes		Ergonomic qualities of keyboards must be ensured via standards.
1990	Stewart		European Directive regarding VDUs.

(Continued)

TABLE C.5 (*Continued*)
Publications from 1981 to 1990 Related to Cumulative Trauma Disorders, Keyboarding, and Keyboard Design

Year	Author(s)	Cumulative Trauma Disorder Aspects	Keying and Keyboard Design
1990	Lachnit and Pieper		New data on speed of finger motions.
1990	Ayoub	CTDs associated with extreme postures, excessive force, concentration of stress, and static loading.	Bank encoding console enforces unnatural and extreme postures of operator.
1990	Hadler	CTDs an iatrogenic concept.	
1990	Council of the European Communities		European Directive including keyboards that can be arranged to avoid fatigue and keyboards that offer support for hands and arms.
1990	Helme et al.	Changes in pain sensitivity and psychometric measures of CTD sufferers.	
1990	Low	Development of CTD symptoms related to duration of keyboarding.	Duration of work with keyboards related to development of symptoms.
1990	Thomson		In a variable-geometry keyboard, an 18° slant and 30°–60° lateral tilt minimized EMG activities and subjective discomfort.
1990	Heyer et al.	VDT operators had higher prevalence of musculoskeletal symptoms than non-VDT operators.	
1990	Burt (NIOSH)	Association of CTDs and keyboarding.	CTDs associated with keyboarding.
1990	IBM	Cumulative strain disorders may occur at VDTs.	Guide to answer questions about radiation, vision, cumulative strain disorders, stress, and ergonomics at VDTs.
1990	Guggenbuehl and Krueger	If the natural keying rhythm cannot be maintained, the musculoskeletal load increases.	The force characteristics of the keys must suit the motor programs to avoid high musculoskeletal loads.

Source: Adapted from Kroemer, K.H.E., *Int. J. Universal Access Inform. Soc.*, 1(2), 99, 2001.

TABLE C.6

Publications from 1991 to 2000 Related to Cumulative Trauma Disorders, Keyboarding, and Keyboard Design

Year	Author(s)	Cumulative Trauma Disorder Aspects	Keying and Keyboard Design
1991	Center for Office Technology		Review of Arndt (1983): *Design for Workers.*
1991	Cushman and Rosenberg		Key and keyboard design guidelines.
1991	Rempel and Gerson		Actual fingertip forces influenced by force–displacement characteristics of keys.
1991a	Kroemer		Ternary chord keyboard a possible competitor for QWERTY keyboard.
1991	Draganova	In Bulgaria, CTD complaints were made by 78% of data entry operators.	
1991	Caple and Betts	Various factors influenced the Australian RSI epidemic.	
1991	Hahn et al.	CTD reduction by training and education, medical treatment, job design and placement, and workstation design.	
1991	Armstrong et al.	Intracarpal canal pressure measured with hand tasks may be sufficient to affect the median nerve.	
1991	Haegg	Low-level muscle loads can affect muscle disorders depending on duration.	
1991	Moore et al.	Hand CTDs explained in terms of external and internal demands, particularly regarding force, movement, repetition, and duration.	
1991	Armstrong	History, causes, pathomechanics, and control program for CTDs.	
1991	Rose		Finger force applied to keys depends on hand–wrist support.
1991b	Kroemer	Case studies and anecdotal evidence for CTD among keyboard operators.	
1991	Apple Computer	CTD may occur by muscle or tendon overuse related to posture, repetition, and force while using computers.	Guidelines on how to place keyboard, hold wrists, use light touch, and take breaks.

(Continued)

TABLE C.6 (*Continued*)
Publications from 1991 to 2000 Related to Cumulative Trauma Disorders, Keyboarding, and Keyboard Design

Year	Author(s)	Cumulative Trauma Disorder Aspects	Keying and Keyboard Design
1991	Stock	Strong evidence for causal relationship between repetitive forceful work and musculoskeletal disorders.	
1991	Sauter et al.		Self-reported data on musculoskeletal discomfort were collected from more than 900 VDT users, and worker posture and workstation were measured in 40 of these users. Effects of ergonomic features on musculoskeletal discomfort were clearly evident from the evaluations. Arm discomfort was associated with high keyboards.
1992	Akagi		Key resistances were not associated with any significant typing speed differences. The highly resistant linear keyboard was the least liked but showed the fewest errors.
1992	Hadler	While repetitive use of the upper extremity can alter hand structure, this does not lead to increased prevalence of any musculoskeletal disease nor to carpal tunnel syndrome. However, repetitive use can lead to soreness. Arm discomfort is real. But psychosocial aspects overwhelm workstation design aspects.	
1992	Lee et al.	An evaluation of 127 exercises recommended for prevention of musculoskeletal discomfort among VDT office workers. Some exercises posed potential safety hazards, exacerbated biomechanical stresses common to VDT work, or contraindicated for individuals with health problems.	

(Continued)

TABLE C.6 (*Continued*)
Publications from 1991 to 2000 Related to Cumulative Trauma Disorders, Keyboarding, and Keyboard Design

Year	Author(s)	Cumulative Trauma Disorder Aspects	Keying and Keyboard Design
1992	Bernard et al. (NIOSH)	Evidence is provided that increasing time spent on computer keyboards is related to the occurrence of CTDs particularly for symptoms and physical findings in the hand/wrist area. For the hand and wrist, psychosocial variables were not as strong predictors as the job task variables.	
1992	Grant		Patent of a keyboard that can be moved along the face of a display. The keyboard is split; the halves slanted and tilted down to the sides. The keys at the side ends of the keyboard and the cursor control keys are designed as mnemonic icons.
1992	Flanders and Soechting		When typing a single letter, all fingers of the hand and the wrist are in motion, in a highly repeatable pattern. The simultaneous motion of the hands is not related.
1992	Soechting and Flanders		Typing is serially executed, letter by letter. When consecutive motions are done with different fingers, the motions can overlap. Movement planning encompasses strings of letters.
1992a	Rempel et al.	Pathophysiology, epidemiology, clinical evaluation, medical management, and prevention of work-related CTDs.	
1992b	Rempel et al.		A new technique is available to measure the fingertip force during key stroke that has three distinct phases: switch compression, finger impact, and pulp compression.

(Continued)

TABLE C.6 (*Continued*)

Publications from 1991 to 2000 Related to Cumulative Trauma Disorders, Keyboarding, and Keyboard Design

Year	Author(s)	Cumulative Trauma Disorder Aspects	Keying and Keyboard Design
1992	VDT News	Various aspects of CTDs are related to physical and psychophysical conditions.	Several articles are included that describe existing "ergonomic" keyboards.
1993	McAtamney and Corlett	Rapid upper limb assessment is a survey method for ergonomic investigations of workplaces where work-related upper limb disorders are reported.	
1993	Schoenmarklin and Marrras	Position, angular velocity, and angular acceleration of the wrists of 40 industrial workers who performed highly repetitive and hand-intensive jobs of high and low risks for CTD were measured. Acceleration best predicted membership to either the high- or low-risk group.	
1993	Ranney	It is important to first identify the tissue injured, then the nature of pathology, and finally the cause. Muscles may show tears or fatigue, tendons microtears and synovial thickening, and nerves hypoxia. The author describes specific steps for diagnosis and procedures to be undertaken for prevention and treatment.	
1993	Wright	New translation of B. Ramazzini's 1713 "De Morbis Articum."	
1993	Angelaki and Soechting		Keystroke kinematics of the hand and fingers usually, but not always, vary with the typing rate. The variations in motions were mostly in the period before the key is hit.
1993	Pascarelli and Kella	53 injured computer users typically had dorsiflexion of the wrist with the fingers arched, used index and middle fingers to strike keys, and hit the keys hard, each habit associated with specific health deficiencies.	The flat keyboard encourages users to place the wrists on desk or table, often in dorsiflexion and/or ulnar deviation. Strain on forearm and hand muscles, especially when splaying the fingers to reach far keys, starts a cascade of events leading to muscle damage and tendinitis.

(Continued)

TABLE C.6 (*Continued*)

Publications from 1991 to 2000 Related to Cumulative Trauma Disorders, Keyboarding, and Keyboard Design

Year	Author(s)	Cumulative Trauma Disorder Aspects	Keying and Keyboard Design
1993	Keyserling et al.	Checklist to determine the presence of ergonomic risk factors in repetitiveness, local mechanical contact stress, forceful manual exertions, awkward upper extremity posture, and hand–tool use.	
1993	Quill and Biers		In an on-screen keyboard, the one-line alphabetic arrangement of keys was superior to the three-line QWERTY layout. Use of a mouse was faster than using arrow keys.
1994	Smutz et al.		A new measuring system to determine the effectiveness of alternative keyboards on finger force, wrist posture, operator comfort, and keying performance.
1994	Carter and Banister	Possible causes of musculoskeletal pain related to the design of workstation, chair, and keyboard and operator selection, training, conditioning, posture, and rest breaks.	Possible causes of musculoskeletal pain related to the design of the keyboard.
1994	Wells et al.	A new system to measure the risk factors for work-related musculoskeletal disorders by combining a video image with quantitative risk information.	
1994	Fernstroem et al.	EMG activities in six forearm and shoulder muscles were higher when using a mechanical keyboard than when using electromechanical and electronic keyboards.	EMG activities in six forearm and shoulder muscles were higher when using a mechanical keyboard than when using electromechanical and electronic keyboards.
1994	Hales et al.	In a cross-sectional study of VDT users 533, the hand–wrist area was most often affected by musculoskeletal disorders. Tendon-related disorders were most frequent, followed by nerve entrapment syndromes. Psychosocial factors play a role in the occurrence.	

(*Continued*)

TABLE C.6 (*Continued*)
Publications from 1991 to 2000 Related to Cumulative Trauma Disorders, Keyboarding, and Keyboard Design

Year	Author(s)	Cumulative Trauma Disorder Aspects	Keying and Keyboard Design
1994	Gerard et al.	Compared to an IBM PS/2 keyboard, muscular activities in users were reduced on a Kinesis keyboard.	Compared to an IBM PS/2 keyboard, muscular activities in users were reduced on a Kinesis keyboard. Learning to achieve high performance was fast on the Kinesis model.
1994	Armstrong et al.		The measured peak forces actually exerted to keys were 2.5–3.9 times the forces required to activate the keys.
1994	Martin et al.		Estimated peak typing forces averaged about 4.6 times the needed key activation force.
1994	Rempel et al.		Typing with keys that had a "make" force of 0.28 and 0.56 N did not affect the applied finger tip force or finger flexor EMG. Both increased when the make force was 0.83 N.
1994	Kroemer		Discussion of the shortcomings of traditional keyboards for current computer use and of possible better ways to transfer information from the human to the computer.
1994	McMulkin and Kroemer		Five subjects learned to operate a one-hand ternary chord keyboard with 18 characters. After about 60 hours of practice, their average keying was at 170 characters per minute.
1994	Rudakewych et al.		A negative keyboard slope significantly improved the posture of the hand, wrist, forearm, and upper arm and sitting posture in all 19 subjects.

(*Continued*)

TABLE C.6 (*Continued*)
Publications from 1991 to 2000 Related to Cumulative Trauma Disorders,
Keyboarding, and Keyboard Design

Year	Author(s)	Cumulative Trauma Disorder Aspects	Keying and Keyboard Design
1994a	Rempel and Horie	Increasing wrist extension deviations from zero while typing produced increasing pressure in the carpal tunnel.	
1994b	Rempel and Horie	Resting the wrist either on a wrist rest or on the table while typing showed significantly increased pressure in the carpal tunnel vis-a-vis keeping the hand "floating" over the keyboard.	
1994	Owen	From the U.S. legal perspective, CTS is preventable, both on and off the job. Administrative and engineering controls can reduce hazardous exposure. People should be kept productive and healthful by fitting their activities to them.	
1994	Burastero et al.		A methodology to compare the results of biomechanical and performance data and subjective assessments during use of various keyboards.
1994	Cakir		A split and adjustable keyboard was compared to a standard keyboard both in a laboratory study and in a 6-month field study. The ergonomic design can improve postural comfort and well-being and reduce fatigue.
1994	Faucett and Rempel	Musculoskeletal problems of VDT operators were frequent. The severity of the problems was associated with work posture, job characteristics, and length of exposure. Psychosocial factors at work affected the symptoms.	
1994	Marras	VDT-related CTD concerns and their prevention by ergonomics measures.	

(Continued)

TABLE C.6 (*Continued*)

Publications from 1991 to 2000 Related to Cumulative Trauma Disorders, Keyboarding, and Keyboard Design

Year	Author(s)	Cumulative Trauma Disorder Aspects	Keying and Keyboard Design
1994	Rempel et al.		During moving and pointing, mean pinch forces applied to an Apple computer mouse were about 0.5 N, but 1.4 N while dragging and almost 4 N when lifting and dragging.
1995	Grant		A patent of a split keyboard, the halves slanted and tilted down to the sides; the center-mounted space bars can be activated in several directions. The keyboard sections can be set to different slope angles.
1995	Moore and Garg	Six task descriptors are rated, and then multipliers are assigned to each. The product of the multipliers is the Strain Index.	
1995	Hedge et al.	Using a keyboard with a negative slope improved the postures of the wrist and body and reduced musculoskeletal discomfort and was strongly preferred, in comparison to a conventional arrangement.	Using a keyboard with a negative slope was strongly preferred in comparison to a conventional arrangement with positive slope.
1995	Masali		The body has the natural tendency to look downward to close targets, with accompanying bends of the neck and trunk, rather than look forward or upward to targets such as a computer screen.
1995	Fogleman and Brogmus	Claim statistics of CTDs shows a growing problem associated with the use of the computer mouse.	
1995	Cakir		Split keyboards with adjustable tilt angles can improve postural and general comfort and reduce fatigue, as shown in short- and long-term studies.

(*Continued*)

TABLE C.6 (*Continued*)
Publications from 1991 to 2000 Related to Cumulative Trauma Disorders, Keyboarding, and Keyboard Design

Year	Author(s)	Cumulative Trauma Disorder Aspects	Keying and Keyboard Design
1995	Yoshitake		The center-to-center distance between keys can be reduced from 19 to 16.7 mm for fast typists even with large fingers and to 15 mm for small fingers, without reducing performance.
1995	Hedge and Powers		Subjects typed on a traditional 101-key keyboard placed on a 68 cm high horizontal surface (a) with and (b) without arm support and (c) with the keyboard placed at a negative slope of 12°. Ulnar deviation did not change, but dorsal wrist extension was 13° in conditions (a) and (b) and reduced to −1° with (c).
1995	Hoffmann et al.		On simulated keyboards with different key sizes and key spacing, the subjects' movement times were shortest when the spacing approximated the size of the finger pad.
1995a	Bergqvist et al.	Among 353 office workers, the occurrence of muscle problems was not different in VDT and non-VDT workers. However, the combination of VDT work of more than 20 hours per week and factors such as limited rest breaks and repetitive movements was associated with risk.	
1995b	Bergqvist et al.	A study of 260 VDT office workers identified factors associated with the occurrence of musculoskeletal problems. Among the organizational factors were flexible rest breaks, task flexibility, and overtime. Among the ergonomic factors were static work posture, hand position, repetitive movements, and keyboard and VDT vertical position.	

(Continued)

TABLE C.6 (*Continued*)

Publications from 1991 to 2000 Related to Cumulative Trauma Disorders, Keyboarding, and Keyboard Design

Year	Author(s)	Cumulative Trauma Disorder Aspects	Keying and Keyboard Design
1996	Honan et al.		Over 4 hours of intensive typing, wrist posture did not change in users of a split and tilted keyboard, but did so in users of a conventional keyboard.
1996	Habes	The author defines and describes CTDs, their development due to activities on and off the job, their incidence rates and costs, and ergonomic means to control them.	
1996	Sommerich et al.		When typing at preferred speed, there is no relationship between speed and key strike force; but if the typing speed is affected by external factors, strike forces tend to increase or decrease with similar changes in typing speed.
1996	Martin et al.	Relatively poor correspondence between EMG data and individual dynamic finger forces may result from the fact that actual muscle load is higher than reflected in key strike force.	Average peak typing forces were about 10% of MVC and 5.4 times the needed key activation force.
1996	Marklin and Simoneau	Split and inclined keyboards appear to reduce wrist ulnar deviation and forearm pronation from that on the conventional keyboard.	Split and inclined keyboards appear to reduce wrist ulnar deviation and forearm pronation from that on the conventional keyboard.
1997	Olecsi and Beaton		A new technique for continuous force–displacement measurement that assesses dynamic force applied to a key.
1997	Straker et al.	Significantly, greater neck flexion and head tilt occurred during work with laptop as compared to desktop computers, but other body angles showed no differences.	Keying performance showed no differences between laptop and desktop computers.

(Continued)

TABLE C.6 (*Continued*)
Publications from 1991 to 2000 Related to Cumulative Trauma Disorders, Keyboarding, and Keyboard Design

Year	Author(s)	Cumulative Trauma Disorder Aspects	Keying and Keyboard Design
1997	Swanson et al.		After 2 days of keying on split, tilted, and sloped keyboards, 50 subjects showed no significant differences in discomfort, fatigue, or performance compared to work on a conventional keyboard.
1997	Rempel et al.	Increasing forces exerted with the fingertip from 0 to 12 N increased carpal tunnel pressure independently from various wrist flexion and extension angles.	
1998	Matias and Salvendy	The main causes of CTS are long periods of continuous typing, static and bent wrist postures, seating posture, and wrist size.	
1998	Martin et al.		Intramuscular and surface EMGs are in good agreement; the flexor carpi radialis and ulnaris muscles are the prime movers in typing on flat keyboards; muscle load is twice as high in extensors than in flexors, and muscle load increases linearly with typing speed.
1998	Greening and Lynn	Perception of vibration may be useful for early detection of onset of RSI.	
1999	Carayon et al.	There are possible links between stressful psychosocial factors at work and force, repetition, and posture.	
1999	Marklin et al.	Split and inclined keyboards reduced the ulnar deviation from that on the conventional keyboard.	
1999	Tittiranonda et al.	80 computer users with upper extremity musculoskeletal disorders showed improving trends in pain severity and hand function, and more individual satisfaction, when using the split, slanted, and tilted keyboards than with a conventional keyboard.	

Source: Adapted from Kroemer, K.H.E., *Int. J. Universal Access Inform. Soc.*, 1(2), 99, 2001.

REFERENCES

Ayoub, M. A. and Wittels, N. E. (1989). Cumulative trauma disorders. *International Reviews of Ergonomics* 2, 217–272.

Banaji, F. M. M. (1920). Keyboard for typewriters. Patent 1,336,122. United States Patent Office, Alexandria, VA.

Burnette, J. T. and Ayoub, M. A. (July–August 1989). Cumulative trauma disorders. Part 1. The problem. *Pain Management* 2, 196–209.

Burry, H. C. and Stoke, J. C. J. (1985). Repetitive strain injury. *New Zealand Medical Journal* 98, 601–602.

Conn, H. R. (September 1931). Tenosynovitis. *The Ohio State Medical Journal* 27, 713–716.

Dvorak, A. (1936). Typewriter keyboard. Patent 2,040,248. United States Patent Office, Alexandria, VA.

Fry, H. J. H. (1986a). Overuse syndrome in musicians—100 years ago. A historical review. *The Medical Journal of Australia* 145, 620–625.

Fry, H. J. H. (1986b). Physical signs in the hands and wrists seen in the overuse injury syndrome of the upper limb. *Australian and New Zealand Journal of Surgery* 56, 47–49.

Fry, H. J. H. (September 27, 1986c). Overuse syndrome in musicians: Prevention and management. *The Lancet* 2, 723–731.

Fry, H. J. H. (1986d). What's in a name? The musician's anthology of misuse. In *Medical Problems of Performing Artists and Incidence of Overuse Syndrome in the Symphony Orchestra*. Philadelphia, PA: Hanley & Belfus, pp. 36–38, 51–55.

Hammer, A. W. (1934). Tenosynovitis. *Medical Record*, October 3, pp. 353–355.

Heidner, F. (1915). Type-writing machine. Letter's Patent 1,138,474. United States Patent Office, Alexandria, VA.

Hochberg, F. H., Leffert, R. D., Heller, M. D., and Merriman, L. (1983). Hand difficulties among musicians. *The Journal of the American Medical Association* 249(14), 1869–1872.

Klockenberg, E. A. (1926). *Rationalization of the Typewriter and Its Operation* (in German). Berlin, Germany: Springer.

Kroemer, K. H. E. (2001). Keyboards and keyboarding. An annotated bibliography of the literature from 1878 to 1999. *International Journal Universal Access in the Information Society* 1(2), 99–160.

Lockwood, A. H. (1989). Medical problems of musicians. *The New England Journal of Medicine* 320(4), 221–227.

Lundervold, A. (1958). Electromyographic investigations during typewriting. *Ergonomics* 1, 226–233.

Lundervold, A. J. S. (1951). Electromyographic investigations of position and manner of working in typewriting. *Acta Physiologica Scandinavica* 24(Suppl. 84), 1–171.

Nelson, W. W. (1920). The improvements in connection with keyboards for typewriters. British Patent 155,446.

Pfeffer, G. B., Gelberman, R. H., Boyes, J. H., and Rydevik, B. (1988). The history of carpal tunnel syndrome. *Journal of Hand Surgery* 13-B, 28–34.

Schroetter, H. (1925). Knowledge of the energy consumption of typewriting (in German). *Pflueger's Archiv fuer die Gesamte Physiologie des Menschen und der Tiere* 207(4), 323–342.

Sholes, C. L. (1878). Improvement in type-writing machines. Letter's Patent 207,559. United States Patent Office, Alexandria, VA.

Wolcott, C. (1920). Keyboards. Letter's Patent 1,342,244. United States Patent Office, Alexandria, VA.

Wright, W. C. (1993). *Diseases of Workers, Translation of B. Ramazzini's 1713 "De Morbis Articum."* Thunder Bay, Ontario, Canada: OH&S Press.

Index